PLTMG: A Software Package for Solving Elliptic Partial Differential Equations

Users' Guide 6.0

Frontiers in Applied Mathematics

Managing Editors
for Practical Computing Series

W.M. Coughran, Jr.
AT&T Bell Laboratories
Murray Hill, New Jersey

Eric Grosse
AT&T Bell Laboratories
Murray Hill, New Jersey

Frontiers in Applied Mathematics is a series that presents new mathematical or computational approaches to significant scientific problems. Beginning with Volume 4, this series reflects a change in both philosophy and format. Each volume focuses on a broad application of general interest to applied mathematicians as well as engineers and other scientists.

This unique series will advance the development of applied mathematics through the rapid publication of short, inexpensive books that lie on the cutting edge of research.

Frontiers in Applied Mathematics

Vol. 1	Ewing, Richard E., The Mathematics of Reservoir Simulation
Vol. 2	Buckmaster, John D., The Mathematics of Combustion
Vol. 3	McCormick, Stephen F., ed., Multigrid Methods
Vol. 4	Coleman, Thomas F. and Van Loan, Charles, Handbook for Matrix Computations
Vol. 5	Grossman, Robert, ed., Symbolic Computation: Applications to Scientific Computing
Vol. 6	McCormick, Stephen F., Multilevel Adaptive Methods for Partial Differential Equations
Vol. 7	Bank, Randolph E., PLTMG: A Software Package for Solving Elliptic Partial Differential Equations. *Users' Guide 6.0*

PLTMG: A Software Package for Solving Elliptic Partial Differential Equations

Users' Guide 6.0

Randolph E. Bank
University of California, San Diego

Society for Industrial and Applied Mathematics
siam
Philadelphia 1990

Library of Congress Cataloging-in-Publication Data

Bank, Randolph E., 1949-
A software package for solving elliptic partial differential equations : users' guide 6.0 / Randolph E. Bank.
p. cm. – (Frontiers in applied mathematics ; vol. 7)
Includes bibliographical references (p.) and index.
ISBN 0-89871-255-6
1. PLTMG (Computer program) 2. Differential equations, Elliptic--Numerical solutions–Data processing. I. Title II. Series: Frontiers in applied mathematics : 7.
QA377.B26 1990
515'. 353–dc20 90-9851
CIP

To Barbara

Contents

Preface

Availability and Installation

The source code for *PLTMG* is available from the *Netlib* software distribution network. It is possible that the details of the procedure to obtain the code from *Netlib* will vary with time. Thus, one should begin with an electronic mail request to *Netlib* of the following form:

mail netlib@research.att.com
send index for pltmg

This index file will contain instructions for getting the rest of the *PLTMG* source code by electronic mail or ftp, and will be updated as required. Because of the large size of the source code, ftp should be preferred to electronic mail. For more information on *Netlib*, make an electronic mail request of the form:

mail netlib@research.att.com
send index

or see [20]. The source code available from *Netlib* will be updated as necessary to correct bugs and other errors.

The source code is contained in a number of files. The contents of several of these files are summarized in table 0.1. Most of the files are portable FORTRAN and can simply be compiled and placed in a library. There are five machine dependent routines, namely *GETCMD*, *PLTUTL*, *PLINE*, *PFILL*, and *TIMER*; these routines may have to be modified for your local environment (see chapter 10). Also included are graphics drivers for the X-Window

File Name	Contents
pltmg.f, etc.	*PLTMG* and associated routines
trigen.f, etc.	*TRIGEN* and associated routines
triplt.f, etc.	*TRIPLT* and associated routines
gphplt.f, etc.	*GPHPLT* and associated routines
inplt.f, etc.	*INPLT* and associated routines
pltevl.f, etc.	*PLTEVL* and associated routines
skeltn.f, etc.	*SKELTN* and associated routines
xgraph.c	X-Windows graphics interface
psgraph.f	PostScript graphics interface
atest.f	test driver *ATEST* and associated routines
circle.f	routines for test problem *CIRCLE*
square.f	routines for test problem *SQUARE*
domains.f	routines for test problem *DOMAINS*
naca.f	routines for test problem *NACA*
jcn.f	routines for test problem *JCN*

Table 0.1.
PLTMG source code files.

System[1] color workstation (xgraph.c), and a PostScript[2] driven laser printer (psgraph.f). Once these machine dependent subroutines have been successfully modified, they can also be compiled once and placed in the library.

The file atest.f contains the test driver *ATEST* and some associated routines. The remaining files contain the data sets for five different test problems (see chapter 12). To run a particular test problem, the data set file for that problem and atest.f should be compiled and linked with the library containing the rest of the object files. One can create new problem data set files, according to the specifications given in chapter 9, and substitute them for the provided test problems. All variable storage allocation and default parameter settings are made in *ATEST*, and thus can be easily modified to suit the computing environment (see chapter 9).

[1] X-Window System is a trademark of the Massachusetts Institute of Technology.
[2] PostScript is a trademark of Adobe Systems, Inc.

Getting Started

This book is intended mainly for use as a reference manual for the *PLTMG* software package, and does not contain much of the mathematical theory that serves as the theoretical foundation for the code. Brief descriptions of some of the main algorithms are included, where such a description is required to clarify the meaning of input parameters, and the resulting behavior of the code. The references list papers that explain and analyze in more theoretical terms most of the algorithms that are incorporated in the code.

PLTMG itself was developed as a prototype program for testing and verifying the properties of various multigrid and adaptive local mesh refinement algorithms. It was the desire to have a framework where such methods could be easily tested on a wide variety of problems that led to its initial development. Since then, the *PLTMG* software has evolved with the author's research interests.

The introduction contains an overview of the entire package, and a simple example demonstrating some of its elementary capabilities. Once the software has been successfully installed, one can become better acquainted with its use by trying the supplied test driver *ATEST* on some of the test problem data sets. To do this, one can refer to chapter 9 for the documentation of *ATEST*, and to chapter 12 for descriptions of the test problems. Chapter 11 illustrates the use of *ATEST* in solving several test problems, including both the commands given *ATEST* and the responses of the various component subroutines.

Eventually, one will require more knowledge of options for particular subroutines. Each of the major subroutines is documented in its own chapter, and one can refer to the appropriate chapters as required. To some extent, these chapters are intended to be read independently of one another. Subroutine *PLTMG* is, of course, the central routine in the package, so that understanding the entire framework demands some initial understanding of this routine.

Acknowledgements

Many people have made significant contributions to this version of *PLTMG*; I am indebted to them all for their help.

The original grid refinement algorithms used in *PLTMG* were derived in 1976 as joint work with Todd Dupont of the University of Chicago. The analysis of the hierarchical basis multigrid iterative procedure embodied in *PLTMG* is joint work with Todd Dupont, and Harry Yserentant of Tübingen University. The data structures and associated algorithms used in *PLTMG* were originally designed and implemented jointly with Andrew H. Sherman, currently of Yale University. The approximate Newton strategies incorporated in the present version of *PLTMG* represent joint work with Donald J. Rose of Duke University. The *a posteriori* error estimation procedures used for adaptive mesh refinement are joint work with Alan Weiser of Rice University. The algorithms used in the pseudo-arclength continuation procedures of *PLTMG* are joint work with Tony Chan of the University of California at Los Angeles, and Hans Mittelmann of Arizona State University. The sparse Gaussian elimination procedures used for the coarse grid solution are joint work with R. Kent Smith of AT&T Bell Laboratories. Many of the improvements in *TRIPLT*, and the unrefinement option in *PLTMG* are joint work with Bruno Welfert of the University of California at San Diego.

Discussions with many users of older versions of *PLTMG* and other interested people have led directly and indirectly to many improvements in the current version. In particular, the computational problems arising from my work with the group at Bell Laboratories have strongly influenced the development of *PLTMG*, *TRIGEN*, *TRIPLT*, and *SKELTN*. I would also like to thank William Coughran and Eric Grosse of Bell Labs for their help in producing this manual and in making the source code available through *Netlib*. Wolfgang Fichtner, Jacques Periaux, R. Bruce Simpson, and R. Kent Smith all provided data that was incorporated into the test problem data sets.

The work of the author on this and previous versions of *PLTMG* has been supported by the Office of Naval Research, currently through contract N00014-89J-1440 (University of California at San Diego). I am especially indebted to Richard Lau for his continuing support.

Chapter 1

Introduction

1.1. Problem Specification

Subroutine *PLTMG* solves boundary value problems of the form

$$-\nabla a(x, y, u, \nabla u, \lambda) + f(x, y, u, \nabla u, \lambda) = 0 \qquad \text{in } \Omega \tag{1.1}$$

with boundary conditions

$$\begin{aligned} u &= g_1(x, y, \lambda) && \text{on } \partial\Omega_1 \\ a{\cdot}n &= g_2(x, y, u, \lambda) && \text{on } \partial\Omega - \partial\Omega_1 \equiv \partial\Omega_2. \end{aligned} \tag{1.2}$$

Here Ω is a connected region in R^2, n is the unit normal, a is the vector $(a_1, a_2)^t$, a_1, a_2, f, g_1, and g_2 are scalar functions, and λ is a scalar continuation parameter. Additionally, the user may specify a functional of the solution of the form

$$\begin{aligned} \rho(u, \lambda) = &\int_\Omega p_1(x, y, u, \nabla u, \lambda)\, dxdy \\ &+ \int_{\partial\Omega} p_2(x, y, u, \nabla u, \lambda)\, ds \end{aligned} \tag{1.3}$$

where p_1 and p_2 are scalar functions. Specification of such a functional is necessary if the problem involves continuation, but is optional otherwise. The software package consists of seven primary subroutines. These main routines and their functions are summarized in table 1.1.

Subroutine *PLTMG* uses finite element discretizations based on C^0 piecewise linear triangular finite elements. The user must provide a crude triangulation of Ω. The data for this triangulation may be directly entered by the user or generated by the input routine

Subroutine	Main Function
PLTMG	solve the partial differential equation (1.1)-(1.3)
TRIGEN	create a coarse grid triangulation of Ω
TRIPLT	display the solution or related function
GPHPLT	display performance statistics for *PLTMG*
INPLT	display the input data defining Ω
PLTEVL	evaluate the solution at an arbitrary point
SKELTN	create a skeleton for Ω based on the solution

Table 1.1.
The main subroutines in the package.

TRIGEN. This crude triangulation may be uniformly and/or adaptively refined by *PLTMG*.

PLTMG uses a continuation procedure to follow solution curves on the coarsest grid. Options are available for continuing to target points in either λ or ρ, finding singular (limit or bifurcation) points, switching branches at bifurcation points, and switching (i.e., defining new) parameters and/or functionals. The user can call for adaptive or user specified mesh refinement at any point along the solution path. There are also options for adaptive unrefinement and subsequent re-refinement of an existing refined mesh, allowing it to adapt to the changing character of the solution along the continuation path. *PLTMG* is described in detail in chapter 2.

Subroutine *TRIGEN* generates an initial mesh from a *skeleton* of Ω. Informally, a skeleton is a description of Ω as the union of a set of non-overlapping subregions, possibly only one. Each subregion is defined in terms of its boundary. *TRIGEN* triangulates these subregions in a consistent fashion, yielding a triangulation of Ω such that the interfaces between subregions are a subset of the edges in the mesh. Subroutine *TRIGEN* is described in detail in chapter 3.

Subroutine *TRIPLT* provides graphical displays of the solution, and other related functions. This graphics package has been extensively revised from previous versions, and now makes extensive use of color. In particular, three dimensional color surface/contour plots with shading and an arbitrary viewing perspective are available. There is also an option for displaying vector functions such as ∇u, in which color indicates direction, and shade indicates magnitude. These and other *TRIPLT* options are described in detail in chapter 4.

Subroutine *GPHPLT* provides a variety of graphical displays of convergence histories, statistical data, and other interesting output from *PLTMG*. Most of this output was not provided in previous versions, or was provided only in tabular form. *GPHPLT* is described in detail in chapter 5. Subroutine *INPLT* provides a graphical display of the input data (triangulation or skeleton) defining Ω. This provides a simple means of visually checking the input data. *INPLT* is described in detail in chapter 6.

Subroutine *PLTEVL* is an output routine designed for evaluating the solution or related function at an arbitrary set of points in Ω. *PLTEVL* is described in chapter 7.

Subroutine *SKELTN* can be used to create an input data set for *TRIGEN* (a skeleton of the region) based on the output of *PLTMG*. For example, one can create by hand a crude skeleton for a given region, use *TRIGEN* to create a triangulation, solve the problem on this grid, and then use *SKELTN* to develop an improved skeleton for the region. In effect, this provides an adaptive rezoning mechanism for the coarsest mesh. *SKELTN* is described in chapter 8.

Each output routine was designed to accept as input the output of *PLTMG* or any of the other output routines. Thus these routines can be called in any order and as frequently as desired. It is also possible to call to *PLTMG* again after calling output routines, allowing one to obtain intermediate output from the calculation if desired.

An elementary interactive test driver, *ATEST*, is described in chapter 9. *ATEST* provides options for calling each of the main routines, as well as other useful functions such as writing and reading data files, resetting parameters, and executing problem specific subroutines provided by the user.

Several short machine dependent routines are required for timing, graphics and file manipulation. These are described in chapter 10. In chapter 11, some example problems are solved using *ATEST*. In chapter 12, the example problem data sets included with the source code are briefly described.

1.2. A Simple Example

In this section, we will present some numerical results for a simple problem solved using *ATEST*. Our intent here is not to provide a detailed description of the particular command sequences we used in solving the problem, but merely to present the results themselves to illustrate some of the elementary capabilities of the package. The use of *ATEST* is fully documented in chapter 9, and more advanced examples, including the relevant *ATEST* command sequences, are given in chapter 11.

In this example, we solve the Poisson equation

$$-\Delta u = 1$$

in Ω, where Ω is given in figure 1.1. The boundary conditions are $u = 0$ on the inner circle and $\nabla u \cdot n = 0$ on the outer circle. Since this is a linear problem with no parameter dependence, it does not make use of the continuation options in *PLTMG*, and it does not require the solution of nonlinear algebraic equations.

The description of Ω was provided as a skeleton consisting of 24 subregions. Thus we must call *TRIGEN* to generate an initial triangulation. The use of so many subregions in the skeleton was unnecessary to define such a simple domain, but we created this skeleton to illustrate some features of *TRIGEN*. In particular, the 24 subregions can be grouped into three sets of eight similar subregions. In our input, we specified that *TRIGEN* should triangulate similar regions in a similar fashion. The resulting triangulation produced by *TRIGEN* is shown in figure 1.1.

The triangulation has 180 vertices and 320 triangles. Notice that the interface lines in the skeleton are retained as triangle edges in the triangulation, and that similar regions are triangulated in a similar fashion. The definition of the skeleton data structures and input data, and a complete description of *TRIGEN* is given in chapter 3.

Subroutine *INPLT* was used to produce figure 1.1. We have used gray scale to approximate the use of color. In the skeleton, each subregion is given its own color, while in the triangulation, regions that share the same (user specified) label are given the same color. In our example, three colors were used to color the triangulation. Other options available in *INPLT* are described in chapter 6.

Having produced an initial triangulation, we can now call *PLTMG* to solve the problem. We did this, specifying that *PLTMG*

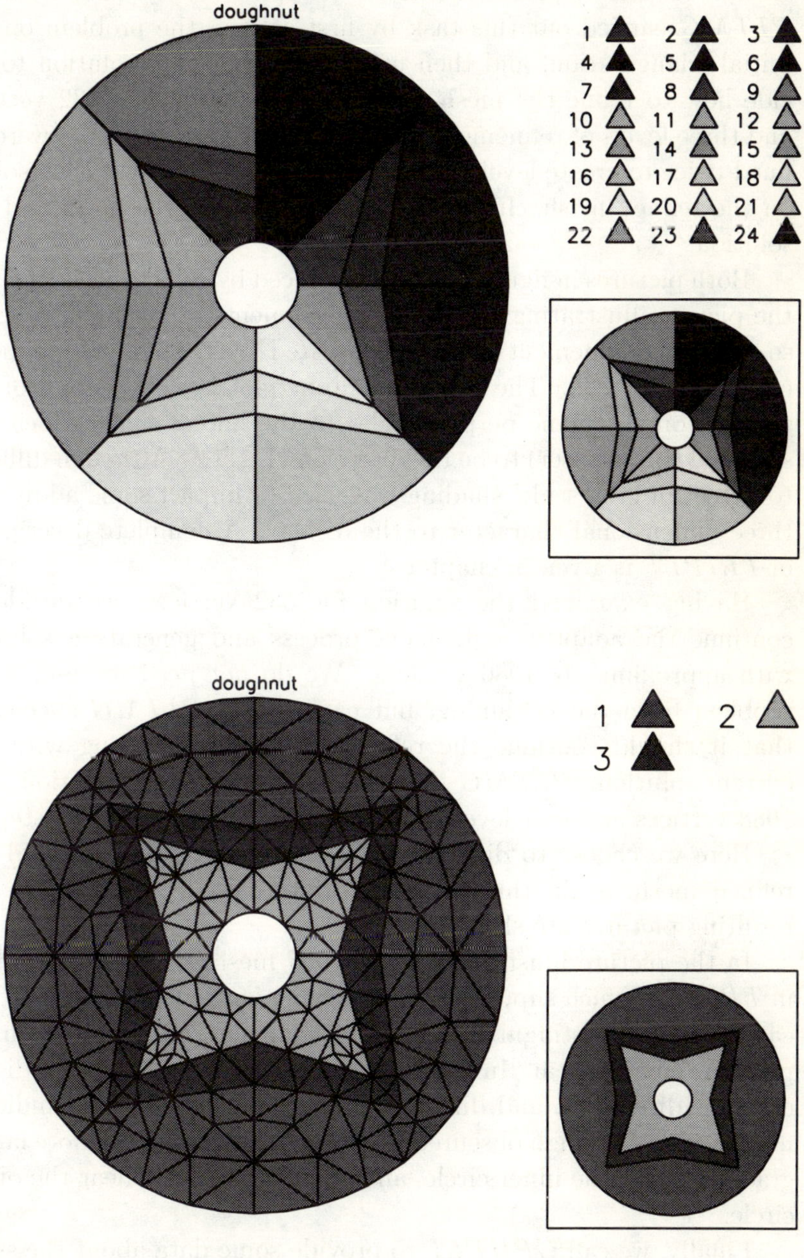

Fig. 1.1. *The skeleton for* Ω *and the triangulation produced by* $TRIGEN$. *Both figures were produced by* $INPLT$.

should adaptively create a mesh with approximately 540 vertices. *PLTMG* carried out this task by first solving the problem on the initial triangulation, and then using that computed solution to decide how to refine the mesh. The resulting mesh had 532 vertices and three levels of refinement (some of the original elements were refined twice to create level 3 triangles). The problem was then solved on the refined mesh. In figure 1.2, we illustrate the mesh and the solution.

Both pictures in figure 1.2 were produced by calls to *TRIPLT*. In the picture illustrating the mesh, the elements are colored by level, so that the refinement pattern generated by the adaptive process can be easily seen. The surface/contour plot is a projection of the solution on the plane perpendicular to the line of sight, which was specified (by the user) to be the direction $(1,1,1)^t$. Although difficult to detect in gray scale, shading was used to impart some additional three dimensional character to the display. A complete description of *TRIPLT* is given in chapter 4.

Having examined the solution for 532 vertices, we decide to continue the adaptive refinement process and generate a solution with approximately 1080 vertices. We do not need to resolve the problem from the beginning, but can reenter *PLTMG* specifying that it should continue the refinement process starting with the current solution. *PLTMG* then is able to generate a solution with 1088 vertices and four levels of refinement in one additional step.

Here we choose to display the gradient of the solution and the refined mesh, again through appropriate calls to *TRIPLT*. The resulting pictures are shown in figure 1.3.

In the picture illustrating the refined mesh, we chose an option in *TRIPLT* which suppresses the drawing of the triangle edges. We can no longer distinguish individual triangles, but the refinement pattern remains clear. In the gradient plot, different colors indicate different directions, and different shades of the same color indicate magnitude. Although obscured by the gray scale, one can note larger gradients near the inner circle, and smaller gradients near the outer circle.

Finally, we call *GPHPLT* to provide some data about the state of the calculation. The result is shown in figure 1.4. On the left are log-log plots of the relative error in $\mathcal{H}^1$ and $\mathcal{L}^2$ norms, as a function of the number of vertices. On a log-log plot, the slope of the curve

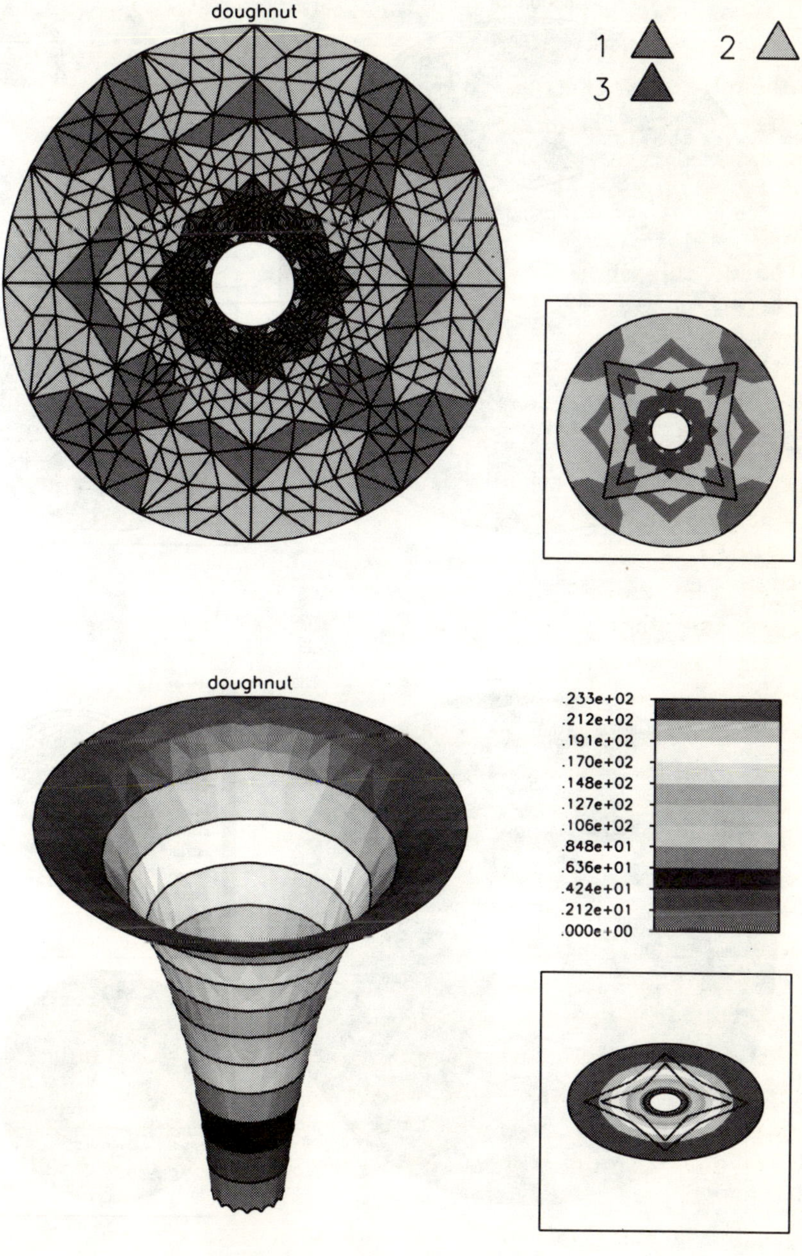

Fig. 1.2. *The refined mesh with 532 vertices and a surface/contour plot of the solution. Both figures were produced by TRIPLT.*

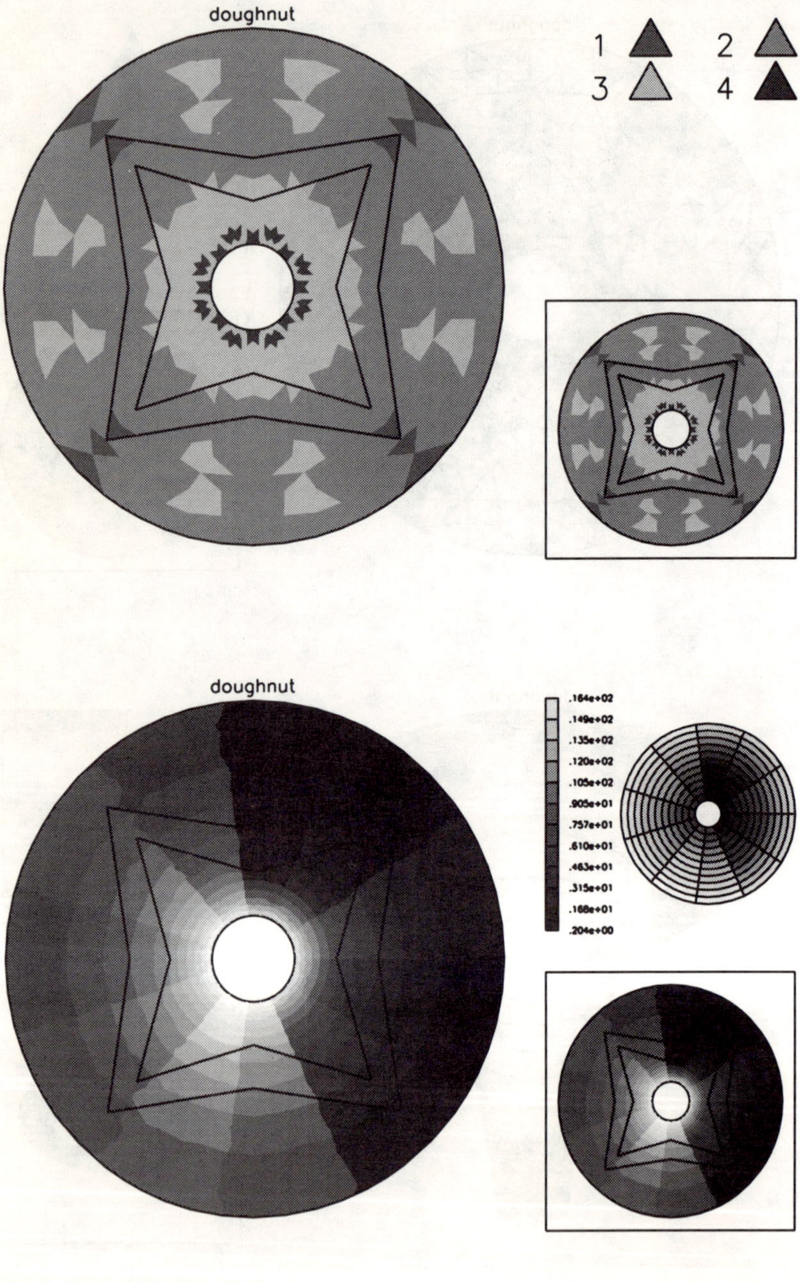

Fig. 1.3. *The refined mesh with 1088 vertices and a plot of the gradient of the solution. Both figures were produced by* $TRIPLT$.

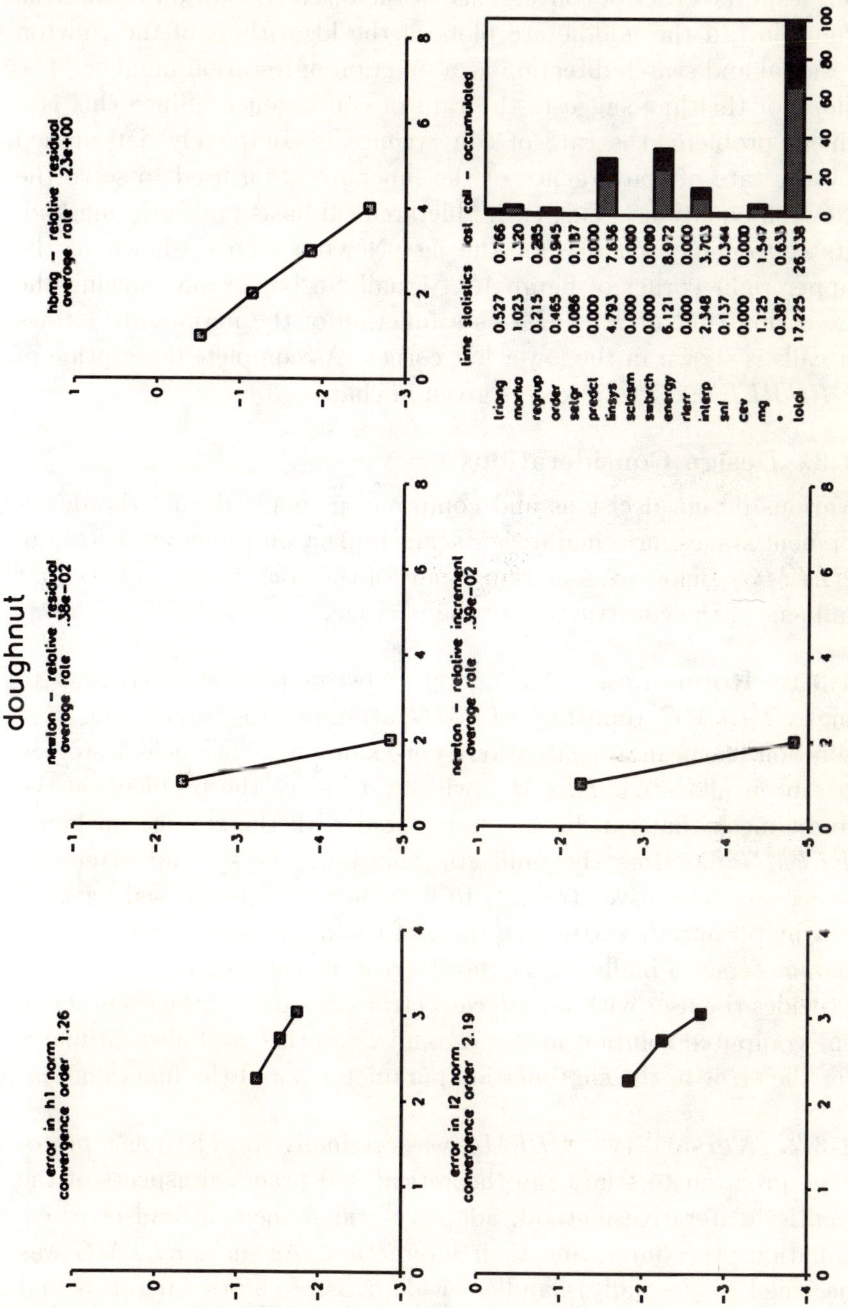

Fig. 1.4. *GPHPLT output.*

suggests the order of convergence of the discrete solution to the true solution. In the middle are plots of the logarithms of the Newton residual and search direction as a function of iteration number. The slope of this line suggests the rate of convergence. Since this is a linear problem, the rate of convergence is completely determined by the rate of convergence of the inner iteration used to solve the Newton equations. This is the hierarchical basis multigrid method. Its convergence history for the last Newton step is shown in the upper right corner of figure 1.4. Finally, a bar graph showing the execution time of *PLTMG* as a function of the major subroutines it calls is shown in the lower left corner. As complete description of *GPHPLT* and its options is given in chapter 5.

1.3. Design Considerations

Various design decisions and compromises made during the development stages have had a significant impact on the current form of *PLTMG*. Below we remark on some of the goals and priorities that influenced the construction of the package.

1.3.1. Robustness Among our most important goals was to make *PLTMG* robust. *PLTMG* attempts to provide the best solution it can in any situation. For example, if insufficient storage has been allocated, *PLTMG* will try to solve the problem on the finest mesh that can be accommodated with the storage on hand. *PLTMG* monitors the multigrid solution process, and attempts, using a conservative strategy, to limit inefficiencies caused by legal but inappropriate settings of some of the more obscure user control parameters. Finally, if no fatal error is encountered, *PLTMG* provides the user with *a posteriori* error estimates of the accuracy of the computed solution in the $\mathcal{H}^1$ and $\mathcal{L}^2$ norms, and also estimates for the error in the continuation parameter λ and the functional ρ.

1.3.2. Versatility *PLTMG* was originally conceived as a prototype program to study the theoretical and practical aspects of the multigrid iterative method, adaptive grid refinement and error estimation procedures, and their interaction. As such, *PLTMG* was designed to (formally) handle a wide class of elliptic operators and reasonably general domains. The boundary of the problem class has expanded as problems were encountered that required its enlargement in order to be solved. The problem class addressed by this version of *PLTMG* should not be interpreted as the limit of the class of problems that could be successfully solved by the techniques embodied by this package. Conversely, one should not assume that every problem (formally) within this class can be solved using the existing code.

This version of *PLTMG* is structured so that it can be reentered any number of times. This allows the user to explore various solution curves by the continuation process returning as required for output and new directives. It is also possible to stop the refinement process at a given level, return for output, and then reenter *PLTMG* and continue the refinement process, possibly with some modifications in strategy .

These capabilities allow considerable flexibility in the use of *PLTMG*, but they also make this version more complicated to use than the earliest editions. In particular, the most important (and, unfortunately, the most complicated) parameters associated with subroutine *PLTMG* are the integers *IPROB* and *IADAPT*. Generally, *IPROB* specifies the task the user is requesting, while *IADAPT* specifies the type of the refinement process to be applied. As $0 \leq IPROB \leq 8$, and $0 \leq IADAPT \leq 4$, there are many possibilities for each parameter, although some combinations of the two are not allowed.

The source code itself is written in a structured fashion, with emphasis on ease of maintenance. For example, the only routines with knowledge of the triangle tree data structures are the routines directly responsible for their creation. All other routines requiring information about triangles and vertices obtain it from a small number of interface routines that retrieve specific types of information from the data structures (the three vertices associated with a given triangle, for example). As another example, there are only five short subroutines that deal directly with the elliptic operator, the boundary conditions, and the functional ρ. These routines assemble element stiffness matrices and right hand sides, and compute various boundary values and integrals on triangle edges. All routines in *PLTMG* which require information about the differential equation obtain it from these routines. It is thus a relatively simple matter to enlarge or contract the problem class, change the quadrature rules, or modify some other aspects of the discretization process.

Finally, we remark that *PLTMG* and the other routines can be used independently of *ATEST*. Indeed, most of the main routines create their own independent internal data structures, so that, after their initialization phases, they become independent of the internal data structures used in *PLTMG*. This has allowed these routines, *TRIGEN*, *TRIPLT* and *SKELTN* in particular, to be easily incorporated into other software packages based on different data structures than those used in *PLTMG*.

1.3.3. Portability *PLTMG* and its associated routines are written in a subset of FORTRAN that should allow them to run with no modification on a wide variety of computers (Apple, CDC, Cray, DEC, Sun, IBM) and operating systems. There are five machine dependent routines (one for *PLTMG* itself, three for the graphics routines *TRIPLT*, *INPLT* and *GPHPLT*, and one for the test driver *ATEST*) that are short, and each carries out elementary tasks.

The routine required by *PLTMG* is for timing. The graphics routines address the graphics device at a primitive level (draw a line, color a polygon, open and close files) and have been used on Tektronix terminals, Sun, DEC, Macintosh, and Apollo workstations, as well as hardcopy devices such as laser printers. The routine required by *ATEST* reads a command line from an input file or device.

1.3.4. Efficiency *PLTMG* is designed for use as an in core program, and makes no direct provisions for secondary storage (although *ATEST* does provide capabilities for saving the solution and associated data sets on disk for later use). Therefore, a premium was placed on data structures and algorithms making efficient use of storage, since we also wanted *PLTMG* to be capable of solving large problems in a wide variety of computing environments. The current triangle tree data structures require four integer storage locations per triangle and four integer storage locations per vertex, plus some additional real and integer information about the user supplied grid. Obtaining information about triangles and vertices usually requires a short computation (the earliest versions of *PLTMG* used more storage and table lookups). The storage allocated for the triangle tree and the solution are the only major storage blocks that remain fixed throughout the package. In particular, storage allocated for matrices and right hand sides in *PLTMG* is used as temporary work space by other routines.

As for time efficiency, during the lifetime of *PLTMG*, this priority has often conflicted with other goals and priorities. Generally, when compromises were made, time efficiency was sacrificed for other objectives. For any given problem, *PLTMG* is probably not the most time efficient program that could be used to solve it, if CPU time is the measure of efficiency. The *a posteriori* error estimates, 3-point quadrature rules, generality of problem class, adaptive refinement procedures, the triangle tree data structures, large number of subroutine calls, and large amount of indirect addressing, among other things, have tended to slow the execution time of *PLTMG*. Generally, *PLTMG* has a reasonable but not spectacular execution time. Our hope is that its flexibility and generality allow it to solve a wide variety of problems quickly in terms of human time.

Chapter 2

Subroutine PLTMG

Subroutine PLTMG is entered using the statement:

Call PLTMG(VX, VY, XM, YM, ITNODE, IBNDRY, A1XY, A2XY, FXY, GXY, P1XY, P2XY, RTRGT, RLTRGT, IP, W)

In subsequent sections, we will explain these parameters and some important features of our implementation.

2.1. The Domain and the Boundary

The domain Ω is specified by providing a crude triangulation. Let $\mathcal{T}$ denote this user triangulation, consisting of NT triangles $\{t_i\}_{i=1}^{NT}$. Triangles may have one curved edge if that edge is part of $\partial\Omega$. Let the triangulation $\mathcal{T}$ have NV vertices $\{v_i\}_{i=1}^{NV}$ and NB boundary edges $\{b_i\}_{i=1}^{NB}$. Also, let $\mathcal{T}$ have NC curved edges with midpoints $\{m_i\}_{i=1}^{NC}$. These are a subset of the NB boundary edges. As an example, consider the circle of radius one with a crack along the positive x-axis. This domain can be triangulated, for example, using 8 triangles, 10 vertices, and 10 boundary edges, 8 of which are curved, as illustrated in figure 2.1.

Vertices v_2 and v_{10} have the same (x, y) coordinates, but v_2 is "above" the crack, and v_{10} "below". Similarly, edge b_1 is the top of the crack, while edge b_{10} is the bottom. The ordering of vertices, triangles, and boundary edges is arbitrary.

The arrays VX and VY are of length NV, and contain as their I-th entries the x and y coordinates of v_I, as illustrated for our example in table 2.1.

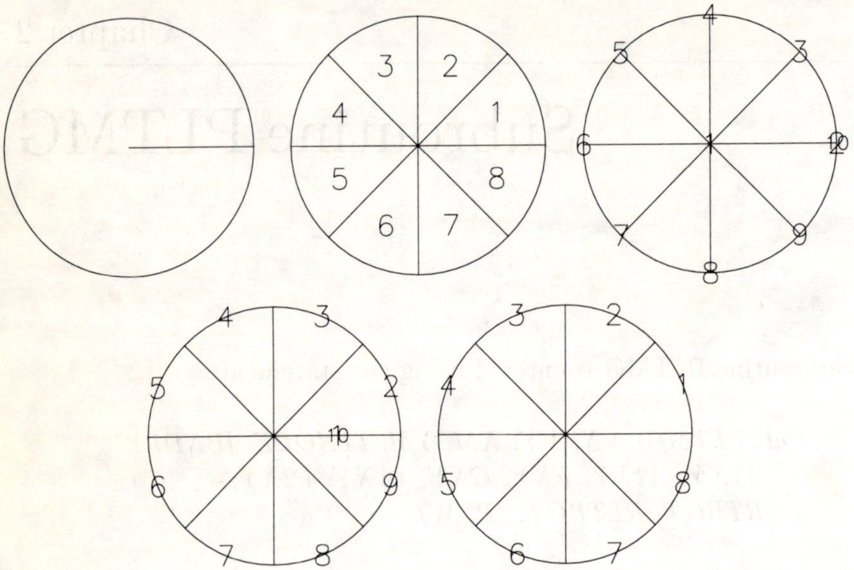

Fig. 2.1. *Clockwise, from upper left: example domain; triangle numbers; vertex numbers; midpoint numbers; boundary edge numbers.*

I	$VX(I)$	$VY(I)$	$XM(I)$	$YM(I)$
1	0	0	$\cos(\pi/8)$	$\sin(\pi/8)$
2	1	0	$\sin(\pi/8)$	$\cos(\pi/8)$
3	$1/\sqrt{2}$	$1/\sqrt{2}$	$-\sin(\pi/8)$	$\cos(\pi/8)$
4	0	1	$-\cos(\pi/8)$	$\sin(\pi/8)$
5	$-1/\sqrt{2}$	$1/\sqrt{2}$	$-\cos(\pi/8)$	$-\sin(\pi/8)$
6	-1	0	$-\sin(\pi/8)$	$-\cos(\pi/8)$
7	$-1/\sqrt{2}$	$-1/\sqrt{2}$	$\sin(\pi/8)$	$-\cos(\pi/8)$
8	0	-1	$\cos(\pi/8)$	$-\sin(\pi/8)$
9	$1/\sqrt{2}$	$-1/\sqrt{2}$		
10	1	0		

Table 2.1.
The VX, VY, XM, and YM arrays. $NV = 10$ *and* $NC = 8$.

If a triangle has a curved boundary edge, that edge is approximated by a circular arc passing through the endpoints of the edge and the midpoint. The arrays XM and YM are of length NC and contain as their I-th entries the x and y coordinates of the midpoint m_I . The data for our example is shown in table 2.1.

$PLTMG$ does not check that the given midpoints are really the midpoints of their respective curved edges. During initialization, the equation for the circle passing through the two endpoints and the midpoint of each curved edge is computed; all subsequent calculations use this characterization. Thus the "midpoint" need only be a third distinct point somewhere on the arc. A curved edge should subtend less than $\pi/2$ of arc.

A given triangle $t_I \in \mathcal{T}$ is specified by giving an accounting of its three vertices, and by specifying an integer label or tag. Insofar as $PLTMG$ is concerned, the tag is arbitrary, and thus it can be used to identify differing region or material properties associated with that element. We shall label the vertices of a triangle ν_i^j, $1 \leq j \leq 3$ The ordering of the vertices of a given triangle is arbitrary and independent of the other triangles. Since $t_I \in \mathcal{T}$, we have $\nu_i^j = v_k$ for some $1 \leq k \leq NV$.

The array $ITNODE$ is a $4 \times NT$ integer array whose I-th column contains information about t_i. The first three entries of $ITNODE$ contain the 3-vertex numbers of triangle t_I. $ITNODE(J, I) = K$, for $1 \leq J \leq 3$, means $(VX(K), VY(K))$ is the J-th vertex of t_I. The fourth entry of column I of $ITNODE$ contains the label for triangle I. In our example, we choose to label each element by the quadrant in the Euclidean plane in which it lies. The $ITNODE$ array for our example is shown in table 2.2.

I	1	2	3	4	5	6	7	8
$ITNODE(1, I)$	1	1	1	1	1	1	1	1
$ITNODE(2, I)$	2	3	4	5	6	7	8	9
$ITNODE(3, I)$	3	4	5	6	7	8	9	10
$ITNODE(4, I)$	1	1	2	2	3	3	4	4

Table 2.2.
The ITNODE array.

The array $IBNDRY$ is a $5 \times NB$ integer array whose I-th column contains information about b_I. The first two entries of $IBNDRY$ contain the indices of the endpoints of the interval (pointers to the VX and VY arrays). $IBNDRY(J,I) = K$, $1 \leq J \leq 2$, means $(VX(K), VY(K))$ is an endpoint of b_I. The ordering of the two endpoints is arbitrary. The third entry of the I-th column contains the index for the midpoint of the edge (pointer to the XM and YM arrays). $IBNDRY(3,I) = K$ means $(XM(K), YM(K))$ is the midpoint of b_I. If there is no midpoint, $IBNDRY(3,K)$ should be set to 0. The fourth entry contains information about the boundary condition type to be imposed on edge b_I. $IBNDRY(4,I) = 1$ means $b_I \in \partial\Omega_2$ (natural boundary conditions). $IBNDRY(4,I) = -1$ means $b_I \in \partial\Omega_1$ (Dirichlet boundary conditions). The fifth entry in column I of $IBNDRY$ contains an arbitrary integer label for the edge, similar to the fourth entry of $ITNODE$; this entry can be used to uniquely distinguish a boundary edge, or to associate some property with the edge. The $IBNDRY$ array is shown in table 2.3.

I	1	2	3	4	5	6	7	8	9	10
$IBNDRY(1,I)$	1	2	3	4	5	6	7	8	9	10
$IBNDRY(2,I)$	2	3	4	5	6	7	8	9	10	1
$IBNDRY(3,I)$	0	1	2	3	4	5	6	7	8	0
$IBNDRY(4,I)$	−1	−1	−1	−1	−1	−1	−1	−1	−1	1
$IBNDRY(5,I)$	−1	0	0	0	0	0	0	0	0	1

Table 2.3.
The $IBNDRY$ array.

In our example, we impose Dirichlet boundary conditions on the outer boundary of the circle, and also along the top of the crack, and Neumann boundary conditions on the bottom of the crack. The outer boundary of the circle will be labeled 0, the top of the crack −1, and the bottom of the crack 1.

2.2. The Partial Differential Equation

The user provides FORTRAN functions $A1XY$, $A2XY$, FXY, GXY, $P1XY$ and $P2XY$ that are related to the functions a and f in (1.2), the functions g_i in (1.2), and the functional ρ in (1.3).

The functions $A1XY$, $A2XY$, FXY, and $P1XY$, all have identical argument lists.

Function A1XY(X, Y, U, UX, UY, RL, ITAG, ITYPE)
Function A2XY(X, Y, U, UX, UY, RL, ITAG, ITYPE)
Function FXY(X, Y, U, UX, UY, RL, ITAG, ITYPE)
Function P1XY(X, Y, U, UX, UY, RL, ITAG, ITYPE)

All arguments are input parameters; the output is returned as the function value itself. Here (X, Y) are the coordinates of a point in element $t_I \in \overline{\Omega}$. U, UX, UY are the values of the function and gradient at the point (X, Y), and RL is the constant λ in (1.1)-(1.2). $ITAG$ is the label assigned to element t_I that was provided by the user as $ITNODE(4, I)$ (all refined elements contained in t_I inherit its label).

$ITYPE$ is an integer between 1 and 5, specifying the output expected by $PLTMG$. For the function FXY, if $ITYPE = 1$, $PLTMG$ is requesting the value of f in (1.1). If $ITYPE = 2$, FXY should return the value of $\partial f/\partial u$; if $ITYPE = 3$, the value of $\partial f/\partial u_x$; if $ITYPE = 4$, the value of $\partial f/\partial u_y$; and if $ITYPE = 5$, the value of $\partial f/\partial \lambda$. ITYPE has an analogous interpretation for $A1XY$, $A2XY$, and $P1XY$.

For example, if the partial differential equation was

$$-\Delta u - \lambda e^u = 0$$

then the functions $A1XY$, $A2XY$, and FXY would be defined as in table 2.4.

$ITYPE$	1	2	3	4	5
$A1XY$	u_x	0	1	0	0
$A2XY$	u_y	0	0	1	0
FXY	$-\lambda e^u$	$-\lambda e^u$	0	0	$-e^u$

Table 2.4.
Coefficient function definitions.

GXY has argument list

Function GXY(X, Y, U, RL, ITAG, ITYPE)

Here (X, Y) are the coordinates of a point on boundary edge $b_I \in \partial\Omega$, U is the value of the function at the point (X, Y), and RL is the constant λ. $ITAG$ is the label assigned to boundary edge b_I that was provided by the user as $IBNDRY(5, I)$.

$ITYPE$ is an integer between 1 and 6. If $ITYPE = 1$, GXY should return the value of g_2 (natural boundary conditions) in (1.2); if $ITYPE = 2$, the value of $\partial g_2/\partial u$; if $ITYPE = 3$, the value of $\partial g_2/\partial\lambda$. If $ITYPE = 4$, GXY should return the value of g_1 (Dirichlet boundary conditions) in (1.2); if $ITYPE = 5$, the value of $\partial g_1/\partial\lambda$.

Finally, if $ITYPE = 6$, GXY should return an initial guess for the solution at the starting point for the continuation process (just an initial guess if there is no λ dependence). For this case only, $(X, Y) \in \overline{\Omega}$, and $ITAG$ is the label of a (possibly nonunique) element t_I containing the point; the input parameter U is not used.

P2XY has argument list

Function P2XY(X, Y, DX, DY, U, UX, UY,
RL, ITAG, ITYPE)

Here (X, Y) are the coordinates of a point on boundary edge $b_I \in \partial\Omega$, (DX, DY) is the unit outward normal for the edge (note that $(-DY, DX)$ is a unit tangent vector), U, UX, UY are the values of the function and gradient at the point (X, Y), and RL is the constant λ. $ITAG$ is the label assigned to boundary edge b_I, and $ITYPE$ is an integer between 1 and 5, defined analogously to the cases of $A1XY$, $A2XY$, FXY, and $P1XY$.

2.3. Discretization, Numerical Quadrature, and Sparse Matrix Storage

Let $\mathcal{T}$ denote a triangulation of Ω and let $\mathcal{M}$ be the space of C^0 piecewise linear polynomials associated with $\mathcal{T}$. *PLTMG* usually represents such a piecewise polynomial using the standard nodal basis; a function can then be specified by giving its values at the vertices. Let $\mathcal{M}_0$ be the subspace of $\mathcal{M}$ whose elements are zero at the knots of $\mathcal{T}$ lying on $\partial\Omega_1$ (the portion of the boundary with Dirichlet boundary conditions), and let $\mathcal{M}_e$ be the affine space of $\mathcal{M}$ whose elements satisfy the Dirichlet boundary conditions at knots of $\mathcal{T}$ lying on $\partial\Omega_1$. The discrete equations solved by *PLTMG* are:

find $u_h \in \mathcal{M}_e$, and λ_h such that

$$a(u_h, v) = \langle g_2, v\rangle \qquad \text{for all } v \in \mathcal{M}_0 \tag{2.1}$$

where

$$\begin{aligned} a(u_h, v) &= \int_\Omega a(u_h, \nabla u_h, \lambda_h) \cdot \nabla v \\ &\quad + f(u_h, \nabla u_h, \lambda_h) v \, dxdy \\ \langle g_2, v\rangle &= \int_{\partial\Omega_2} g_2(u_h, \lambda_h) v \, ds \end{aligned} \tag{2.2}$$

If the Jacobian is not self-adjoint, some upwinding terms based on the divergence free upwinding scheme of [6] are added.

When the continuation process is used, we use a normalization equation of the form:

$$N(u_h, \lambda_h) = \sigma \tag{2.3}$$

The scalar σ is the steplength. *PLTMG* uses two different normalization equations [7], [22]. Most frequently, we use:

$$N(u, \lambda) = \theta\dot{\rho}_0(\rho - \rho_0) + (2 - \theta)\dot{\lambda}_0(\lambda - \lambda_0) \tag{2.4}$$

Here θ is a parameter selected by *PLTMG*; by choosing θ and σ properly, it is possible to achieve target values in either ρ or λ. The vector (u_0^t, λ_0) is the current solution point and $(\dot{u}_0^t, \dot{\lambda}_0)$ the current unit tangent vector. The scalar $\dot{\rho}$ is defined formally using the chain rule for differentiation:

$$\dot{\rho} = \rho_u \dot{u} + \rho_\lambda \dot{\lambda}$$

The other normalization used in *PLTMG* is based on the pseudo-arclength method, characterized by:

$$N(u, \lambda) = \theta\dot{u}_0(u - u_0) + (2 - \theta)\dot{\lambda}_0(\lambda - \lambda_0) \tag{2.5}$$

The normalization (2.5) is used when (2.4) is not well defined (e.g., if $\dot{\rho}_0 = \dot{\lambda}_0 = 0$). For reasonable choices of ρ, these will be isolated points on the solution manifold, such as symmetry breaking bifurcation points. In such instances, (2.5) is used on a temporary basis until (2.4) is defined again.

There are only five relatively short subroutines (*ELEASM*, *ELENBC*, *ELEDBC*, *ELEUN*, and *ELEBDI*) that have knowledge of the discretization procedure and the form of the differential equation (1.1)-(1.2) and the functional ρ in (1.3). $ELEASM$ is used for assembling stiffness matrices, right hand sides, and the contribution to ρ from $P1XY$ for a single element. $ELENBC$ computes contributions to the natural boundary conditions from a single element edge. $ELEDBC$ evaluates Dirichlet boundary conditions and initial guesses at a single point. $ELEUN$ computes an integral on a single interior element edge that is required for the *a posteriori* error estimates. $ELEBDI$ computes the contribution to ρ from $P2XY$ for a single boundary edge.

These routines are written to accept general quadrature rules. The details of the rule (weights and abscissas) are read from tables (that are initialized in $DATA$ statements) making it easy to change the formulas. Currently $ELEASM$ uses a 3-point rule, while $ELENBC$, $ELEUN$ and $ELEBDI$ use 2-point-Gauss rules. The organization within these routines favors clarity at the expense of efficiency, making it reasonably simple to change some aspects of the problem class and discretization.

Global stiffness matrices are stored in the sparse matrix format described in [16]. All nonzeros are stored in a linear array; first the diagonal entries are stored, followed by the upper triangular entries, stored column by column. If the matrix is nonsymmetric, this is followed by the lower triangular entries, stored row by row.

2.4. Level-1 Continuation

We characterize solutions of (1.1)-(1.3) in terms of curves in the (λ, ρ) plane; typical curves are shown in figure 2.2.

The singular point labeled "A" in the figure on the left is a limit (turning) point and those labeled "B" in the figure on the right are bifurcation points (this figure corresponds to the special case of a linear eigenvalue problem).

The purpose of the continuation process is to compute solutions (u, λ) corresponding to points on these curves. The continuation process is carried out entirely on the coarsest mesh, denoted the *level-1* mesh. Note that the definition of the level-1 mesh may change as the calculation proceeds. There are five options for level-1 continuation, characterized by settings of the integer parameter $IPROB$. These are summarized in table 2.5.

An initial solution is provided by the user through function GXY (with $ITYPE = 6$). Thereafter, the continuation proceeds from

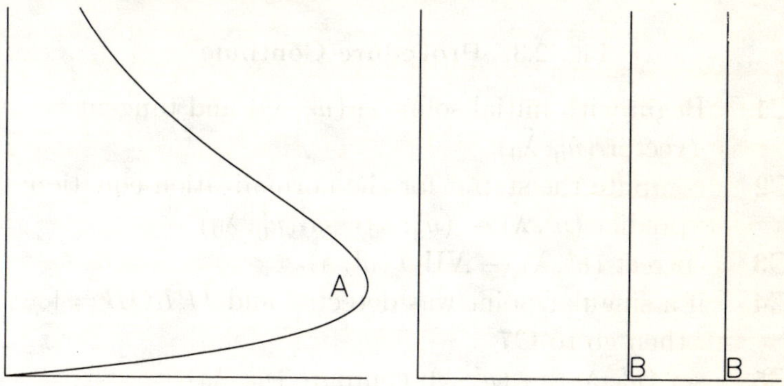

Fig. 2.2. *Continuation Curves* $\rho = \rho(\lambda)$.

$IPROB$	level-1 continuation option
0	continue to the nearest target point
1	continue to the nearest target point or singular point
2	switch branches at a bifurcation point
3	switch λ and/or ρ; initialize with λ fixed
4	switch λ and/or ρ; initialize with ρ fixed
$IADAPT$	
0	normal
1	first call to $PLTMG$

Table 2.5.
The parameters $IPROB$ *and* $IADAPT$*; see also table 2.6.*

the last successfully computed point. A brief outline of the basic continuation process ($IPROB = 0$ or $IPROB = 1$) is given in figure 2.3.

Fig. 2.3. **Procedure Continue**

C1 Begin with initial solution (u_0^t, λ_0) and tangent vector $(\dot{u}_0^t, \dot{\lambda}_0)$

C2 compute the step σ for the normalization equation; predict $(u^t, \lambda) \leftarrow (u_0^t, \lambda_0) + \alpha(\dot{u}_0^t, \dot{\lambda}_0)$

C3 correct $(u^t, \lambda) \leftarrow NWT(u^t, \lambda)$

C4 if a singular point was detected and $IPROB = 1$, then go to **C7**

C5 set $(u_0^t, \lambda_0) \leftarrow (u^t, \lambda)$; compute $(\dot{u}_0^t, \dot{\lambda}_0)$

C6 if (u_0^t, λ_0) is a target point, then exit; else go to **C2**

C7 compute the singular point using secant/bisection algorithm on σ; exit

To continue with $IPROB = 0$ or $IPROB = 1$, the user specifies target values for ρ ($RTRGT$) and λ ($RLTRGT$). $PLTMG$ then attempts to compute a solution with $\rho = RTRGT$ or $\lambda = RLTRGT$, depending on which target value appears to be closer in terms of arclength on the solution curve. If the current value of ρ is $RTRGT$, then $PLTMG$ seeks a solution with $\lambda = RLTRGT$; similarly, if the current value of λ is $RLTRGT$, then $PLTMG$ seeks a solution with $\rho = RTRGT$.

A step size and a predicted solution are computed on line C2. The predictor is of the standard Euler type commonly used in continuation procedures. The step size calculation is influenced not only by the user request, but also by imposed requirements that the predicted solution be sufficiently accurate. The procedures used in this portion of the calculation are described in detail in [11]. The solution is corrected on line C3. The correction process symbolized by the operator NWT involves the solution of a set on nonlinear equations, and is discussed in greater detail in section 2.6

If $PLTMG$ detects a singular point along the solution curve between the starting point and target point, and $IPROB = 1$, the computation of the target point is abandoned in favor of computation of the singular point. A secant/bisection algorithm is used. More details of these p' .edures can be found in Bank and Chan [7] and the references therein. (We remark that the algorithms in $PLTMG$ were designed to handle only simple limit and bifurcation points). When a target or singular point has been successfully computed, $PLTMG$ returns with ($RLTRGT$, $RTRGT$) set to the current values of (λ, ρ).

If $PLTMG$ is called with $IPROB = 2$ at a bifurcation point, parameters relevant for the continuation procedure are initialized for the bifurcating branch, but the solution itself remains unchanged. In the next call to $PLTMG$ with $IPROB = 0$ or $IPROB = 1$, the continuation procedure will follow the bifurcating branch.

If $PLTMG$ is called with $IPROB = 3$ or $IPROB = 4$, parameters relevant for the continuation procedure are reinitialized using the new parameter or functional; the solution itself remains unchanged. The two cases differ in that either λ or ρ can be held fixed during the reinitialization; for either case it is possible to specify either a new continuation parameter λ, a new functional ρ, or both.

The successful use of the level-1 continuation procedures requires some guidance from the user. For example, it is possible to specify target values that cannot be reached. Also, since singular points are detected by changes in sign of the determinant of the Jacobian, one might fool the singular-point detection algorithm by specifying target values sufficiently far away that two singular points are passed on one step and the determinant will not change sign.

The parameter $IADAPT$ also is relevant to the level-1 continuation procedure. In particular, $IADAPT = 0$ and $IADAPT = 1$ are the available options. $IADAPT = 1$ signals the first call to $PLTMG$; all arrays are initialized and the starting point should be specified through function GXY (the initialization process itself is similar to the case $IPROB = 3$). $IADAPT = 0$ indicates that no initialization is needed, and the continuation procedure should start from the current point.

The parameter $LEVEL1$ defines the coarse mesh for the continuation procedure (a full explanation of this parameter is given in section 2.5). For the moment, we remark that after grid refinement at a given point along the continuation path, $LEVEL1$ can be increased, and the continuation process may then proceed on a more refined mesh. On the other hand, if $LEVEL1 > 1$, it may be decreased and the continuation process will proceed on a less refined mesh.

2.5. Grid Refinement

Using *PLTMG*, one can specify uniform grid refinement, adaptive grid refinement, or a combination consisting of some uniform refinement followed by adaptive refinement.

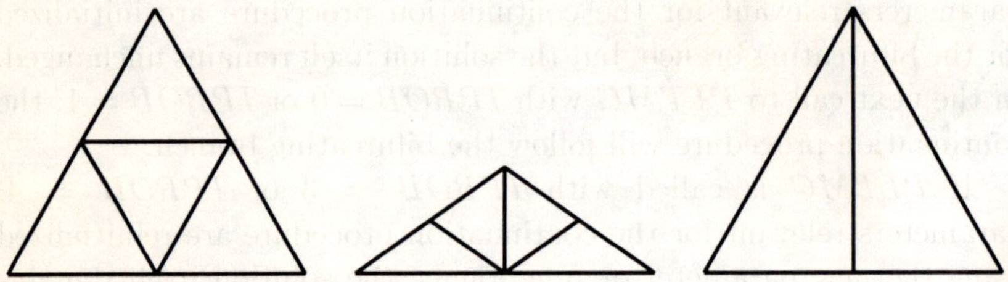

Fig. 2.4. *Regular Refinement (left, center) and Green Refinement (right).*

Normally, a triangle is regularly refined by dividing it into 4 similar triangles, as illustrated in figure 2.4, left. If the triangle has an interior angle greater than $\pi/2$, then it will be regularly refined into two similar triangles, and two geometrically better triangles, as illustrated in figure 2.4, middle. At the boundary of a refined region, it is necessary to divide a triangle into only 2 "green" triangles by inserting a "green" edge, as illustrated in figure 2.4, right. Refinement into green triangles is done only as a temporary measure; at each step of the adaptive procedure, all the green edges are removed and the father elements are considered for regular refinement.

In our refinement procedure, each regularly refined triangle is geometrically similar to or better than the user triangle containing it. When an element is refined into two triangles, the degradation in the geometry is bounded, since green triangles are never further refined. (These remarks are modified slightly, in the obvious fashion, for triangles with a curved edge.) Thus, our refinement procedure can never create triangles with unusually small angles unless the user provides such triangles initially.

The natural data structure associated with our refinement algorithm is a *triangle tree* [15]. This is illustrated schematically in figure 2.5 for a simple two step refinement process (note that one green edge was required to make the mesh a triangulation).

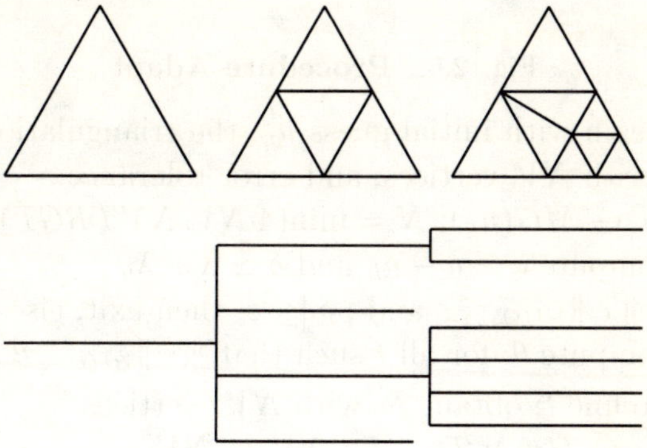

Fig. 2.5. *A Refined Element Tree.*

We define the level ℓ_i of a triangle t_i as follows: $\ell_i = 1$ if t_i is a user triangle; otherwise, $\ell_i = \ell_{f_i} + 1$, where t_{f_i} is the father of t_i on the triangle tree.

Vertices in the mesh are also (uniquely) assigned levels. Vertices of the user provided mesh are level-1 vertices; level-k vertices are those newly created by the refinement of a level-$(k-1)$ triangle. Thus a level-k triangle must have at least one level-k vertex, but may also have one or more vertices of lower level.

Uniform refinement of the mesh is controlled by the parameter $IREFN$. If $IREFN = k$, then the mesh will be uniformly refined to k levels on the triangle tree (e.g., $IREFN = 2$ causes all user elements to be refined once; $IREFN = 3$ causes all elements to be uniformly refined twice, giving 16 refined elements for each user triangle). To specify only adaptive refinement, one should set $IREFN = 1$.

Our adaptive procedure is based on the ideas of Babuška, Rheinboldt, and their co-workers [1], [2], [3], [23], [24], and Bank and Weiser [4], [17], [26]. The procedure used is now commonly called a *feedback* scheme. A brief summary of the procedure is given in figure 2.6; this sketch leaves out many important details, and is intended mainly to illustrate the underlying philosophy of the adaptive process [5].

Line A2 represents the solution of a large set of nonlinear equations; these equations are solved by a damped Newton iteration [13], [12], using the hierarchical basis multigrid iteration to (approx-

Fig. 2.6. **Procedure Adapt**

A1 Begin with initial guess $u_h^{'}$, the triangulation $\mathcal{T}$ with NV vertices, and error tolerance ε

A2 $u_h \leftarrow MG(u_h^{'})$; $N = \min(4{\cdot}NV, NVTRGT)$

A3 compute $e \simeq u - u_h$ and $\delta \simeq \lambda - \lambda_h$

A4 if $\| e \|_{\mathcal{H}^1(\Omega)} \leq \varepsilon$ and $| \delta | \leq \varepsilon$, then exit; else go to **A5**

A5 compute θ; for all t such that $\| e \|_{\mathcal{H}^1(t)} \geq \theta$ refine t; obtain $\mathcal{T}^{'}$ with $NV^{'}$ vertices

A6 $u_h^{'} \leftarrow \mathcal{I}(u_h)$; $\mathcal{T} \leftarrow \mathcal{T}^{'}$; $NV \leftarrow NV^{'}$

A7 if $NV \simeq N$ then go to **A2**; else $u_h \leftarrow u_h^{'}$; go to **A3**

imately) solve the associated linear systems [8], [9], [27],[28]. This aspect of the process will be discussed in more detail in the following section.

For each $t \in \mathcal{T}$, we compute an error estimate by solving a local Neumann problem in t (line A3). The error is approximated in t by a quadratic polynomial that is zero at the vertices of t. In the case where there is no λ dependence, this requires the assembly and solution of a 3×3 set of equations in each element, and leads to a global approximation of the error as a discontinuous piecewise quadratic polynomial that is zero at all vertices of the triangulation.

The local Neumann problems are all linear. The data for the problem in t is the (nonlinear) residual of the partial differential equation in t, and the boundary data is based on the jump in normal direction of the vector function $a(u_h, \nabla u_h, \lambda_h)$ across the edges of t [17], [4].

When a continuation parameter is involved, the error e is still approximated as a discontinuous piecewise quadratic polynomial. However, three different 3×3 linear systems are solved in each element. All involve the same stiffness matrix, but different right hand sides.

On line A5, we refine triangles whose error estimates are above the threshold θ. Often this procedure results in a triangulation $\mathcal{T}^{'}$ in which few new mesh points have been added. However, it is desirable for computational efficiency to have a geometric increase in the dimensions of the problems to be solved; in $PLTMG$ we try to increase the dimension of the space by a factor of 4, or to achieve the target value $NVTRGT$ specified by the user. If this goal is not achieved, the solution is interpolated onto the new mesh, using an interpolation procedure symbolized by the operator $\mathcal{I}$ on line A6, and new error estimates are computed on line A3 without resolving the global problem. Because of the obvious dangers in such a procedure, we use a rather expensive interpolation procedure that updates all values in regions of refinement (not just at newly created vertices) using the linearized differential equation; the effect is something akin to a partial damped Jacobi sweep.

IPROB	refinement option
5	refine, holding the parameter λ_h fixed
6	refine, holding the functional ρ fixed
7	refine on the perpendicular hyperplane to the level-1 solution
8	refine; the problem has no λ dependence
IADAPT	
0	normal
1	first call to *PLTMG*
2	solve on existing meshes
3	continue the refinement process, starting from the existing mesh
4	iterate on the current finest level

Table 2.6.
The parameters IPROB and IADAPT; see also table 2.5.

Two parameters, *IPROB* and *IADAPT*, play important roles in the solution process on refined grids. The use of these parameters is summarized in table 2.5. Also relevant are *NVTRGT*, the (approximate) number of vertices desired in the refined mesh, and *DIGITS*; $10^{-DIGITS} = \varepsilon$ in line A1 of Procedure Adapt.

We begin by noting that the discretization process can introduce spurious solution curves, or cause significant distortions in the solution curves of the continuous problem (1.1); one must therefore be cautious in interpreting the numerical results. As the mesh is refined, the solution curves will generally move; the assumption of *PLTMG* is that, as a function of the discretization, the solution curves converge in some uniform fashion to those of the continuous problem, and that the level-1 mesh is sufficiently fine to capture the qualitative features of the continuous problem's solution curves in the regions of interest [7], [10]. Typically, for each point on the coarse grid, there are three natural points on a nearby fine grid solution curve which might be associated with it: the point with the same λ value, the point with the same ρ value, and the point of intersection with the perpendicular hyperplane passing through the coarse grid solution point. Some typical examples are illustrated in figure 2.7.

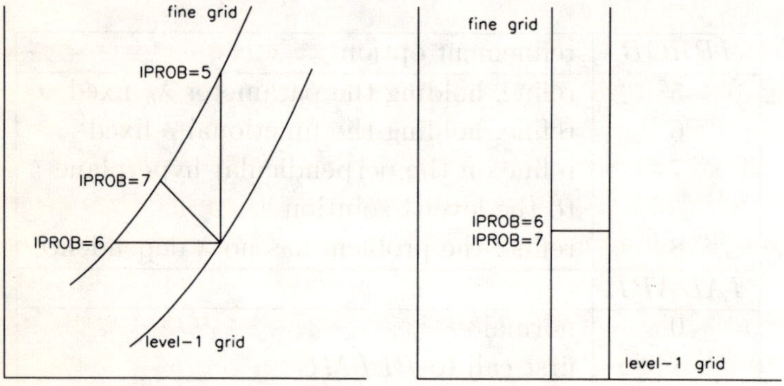

Fig. 2.7. *The effect of IPROB.*

Using $IPROB$, the user can ask $PLTMG$ to compute solutions on refined meshes corresponding to any of these cases. In some situations, all three points may not exist, or they may not be distinct. This is illustrated in figure 2.7, left, where $IPROB = 6$ and $IPROB = 7$ correspond to the same fine grid point, while no (nearby) solution exists for $IPROB = 5$. $IPROB = 8$ specifies there is no scalar parameter λ in the problem, and is used when the continuation features of $PLTMG$ are not desired.

With $IADAPT = 0$ or $IADAPT = 1$, the refinement process is carried out until one of the following conditions is satisfied:

- a refined mesh with approximately $NVTRGT$ vertices is created.
- $10^{-DIGITS}$ of accuracy is achieved in the computed solution.
- storage is exhausted.
- there is a fatal error in the damped Newton iteration.

$IADAPT = 1$ causes initialization to occur, and should be used in this framework only with $IPROB = 8$. For continuation problems, the initialization should occur with $IPROB = 0$; $IADAPT = 0$ should be used for problems in which the level-1 continuation process has been used to generate the coarse grid solution.

If one has previously generated refined meshes, say by adaptively refining the mesh at some previous point in the continuation process, one can ask $PLTMG$ to simply solve the problem on these preexisting meshes. To invoke this option, one should set $IADAPT = 2$. With this option, one can reset the values of $IPROB$ and the iteration parameters $MAXR$, $ITMAX$, and $LEVEL1$.

$IADAPT = 3$ allows $PLTMG$ to return for output during the refinement process. Suppose, for example, we set $NVTRGT = 100$, $IADAPT = 0$, thereby asking $PLTMG$ to create a refined mesh with about 100 vertices. Assuming this is successful, $PLTMG$ will return with the solution computed on that mesh; if we decide the solution is not sufficiently accurate, we can continue the refinement process starting from the current solution, rather than restarting again from the coarse grid (level-1 solution). To do this we can set, for example, $NVTRGT = 250$, $IADAPT = 3$, to create a more refined mesh with about 250 vertices, starting from the mesh with 100 vertices. In the case $IADAPT = 3$, the parameters $DIGITS$, $ITMAX$, $MAXR$, and $LEVEL1$ can be reset, but other parameters should be the same as in the previous call to $PLTMG$.

The option $IADAPT = 3$ also allows for the adaptive unrefinement and/or adjustment of the current mesh. If $PLTMG$ is called with $IADAPT = 3$ and $NVTRGT << NVF$ (NVF is the number of vertices in the current finest mesh), $PLTMG$ will invoke a refinement/unrefinement process yielding a new mesh with (approximately) the required number of vertices. If $PLTMG$ is called with $IADAPT = 3$ and $NVTRGT \sim NVF$, $PLTMG$ will invoke a refinement/unrefinement process that may refine some elements and unrefine others, attempting to produce a better mesh of about the same dimension. This option is useful in continuation problems where the character of the solution changes along the continuation path. After computing the proposed change to the mesh, $PLTMG$ decides whether the change is really significant. If not, $PLTMG$ will return with no adjustment to the mesh.

On the current finest level, the damped Newton or multi-level iteration might fail to converge because of inappropriate settings of $ITMAX$, $LEVEL1$ or $MAXR$. Here we may want to compute a few more damped Newton iterations on the current finest level, starting from the solution as it currently exists on that level. For this option, we specify $IADAPT = 4$, and can reset $ITMAX$, $LEVEL1$ or $MAXR$, but no other parameters.

$PLTMG$ returns with $IADAPT = 0$, and $IPROB = 0$ if called with $2 \leq IPROB \leq 7$. Although convenient in most situations, in some cases it requires the user to reset $IADAPT$ and $IPROB$ to their previous values.

2.6. Equation Solution

The systems of nonlinear equations are solved by a combination of a damped Newton iteration [13], [12] and the hierarchical basis multigrid iteration [9], [8], [18], [27], [28]. When the problem has λ dependence, the bordered system of equations is solved by block Gaussian elimination; this requires the solution of two sets of equations using the multigrid iteration. Suppose the system to be solved has the form:

$$G(u_h, \lambda_h) = 0 \tag{2.6}$$

$$N(u_h, \lambda_h) = \sigma \tag{2.7}$$

Here the operator G represents the finite element equations of order NV, and N the normalizing equation used in the continuation process; σ is the steplength. At any given step of the Newton process, the system to be solved has the form

$$\begin{bmatrix} G_u & G_\lambda \\ N_u & N_\lambda \end{bmatrix} \begin{bmatrix} \epsilon \\ \delta \end{bmatrix} = - \begin{bmatrix} G(u_h, \lambda_h) \\ N(u_h, \lambda_h) - \sigma \end{bmatrix} \tag{2.8}$$

where ϵ is a vector of length NV and δ is a scalar. The solution is constructed by solving

$$\begin{aligned} G_u v &= -G \\ G_u w &= -G_\lambda \\ (N_u w + N_\lambda)\delta &= -(N_u v + N - \sigma) \\ \epsilon &= v + \delta w \end{aligned}$$

The actual implementation involves the deflation technique of Chan [19], allowing the process to remain well-defined even when G_u becomes singular. This requires the computation of the left and right singular vectors of G_u corresponding to its smallest singular value. This is done by an inverse iteration process involving the solution of several sets of equations with G_u and G_u^t.

The block elimination process is embedded in an overall damped Newton process [12] given in figure 2.8. Here $U_k^t = (u_h^t, \lambda_h)$ is the

Fig. 2.8. **Procedure Newton**

N1 Begin with initial guess U_0, and a sufficient decrease parameter τ

N2 $\mathcal{K} \leftarrow 0$; $k \leftarrow 0$; compute $\mathcal{G}_0$ and $\| \mathcal{G}_0 \|$

N3 compute χ_k

N4 $s_k \leftarrow (1 + \mathcal{K} \| \mathcal{G}_k \|)^{-1}$

N5 compute $U_{k+1} = U_k + s_k \chi_k$, $\mathcal{G}_{k+1}$, and $\| \mathcal{G}_{k+1} \|$

N6 if $(1 - \| \mathcal{G}_{k+1} \| / \| \mathcal{G}_k \|) s_k^{-1} < \tau$
then increase $\mathcal{K}$ and go to **N4**;
else decrease $\mathcal{K}$ and set $k \leftarrow k + 1$

N7 if converged, then exit; else go to **N3**

k-th Newton iterate, $\chi_k^t = (\epsilon^t, \delta)$, and $\mathcal{G}_k^t = (G^t, N - \sigma)$. The scalar s_k is the damping parameter, chosen by the formula N4, involving the nonnegative scalar $\mathcal{K}$. The block Gaussian elimination procedure is implicit in the computation of χ_k on line N3. The same damping strategy is also applied when there is no λ dependence, with the obvious redefinition of U_k, χ_k, and $\mathcal{G}_k$. A maximum of $ITMAX$ damped Newton iterations are allowed, where $ITMAX$ is a user specified integer.

All the sets of linear equations involving the matrices G_u and G_u^t have the appearance of finite element discretizations of linear elliptic boundary value problems, and are solved by the hierarchical basis multigrid iterative procedure [9], [18]. Although quite different in its mathematical formulation and analysis, at the level of implementation the procedure is exactly analogous to a classical multigrid v-cycle, except that only a subset of the unknowns are smoothed on each level.

The concept of levels for the hierarchical basis multigrid method and the concept of levels for grid refinement are similar. The parameter $LEVEL1$ defines the coarse grid for purposes of the hierarchical basis multigrid method. If $LEVEL1 = 1$, the coarse grid is the user defined mesh (input mesh for $PLTMG$), and the level-j

mesh consists of all the elements on the triangle tree with $\ell_t = j$, and all elements of level less than j that are unrefined. If $LEVEL1 = m$, $m > 1$, then the level-m mesh with respect to the triangle tree becomes the coarse mesh in the hierarchical basis iteration, and the level-j mesh with respect to the triangle tree becomes the level-$(j-m+1)$ mesh with respect to the hierarchical basis (all the mesh levels are shifted by the constant $LEVEL1$). The $LEVEL1$ mesh plays a distinguished role in the package because sparse Gaussian elimination is used to solve the coarse mesh equations, and because it is the mesh used for the continuation process. A single cycle of the hierarchical basis multigrid method is summarized in figure 2.9.

Fig. 2.9. **Procedure V-Cycle**

V1 Begin with initial guess U, $LEVEL1$, and $LVLF$

V2 for $k = LVLF$ to $LEVEL1 + 1$ in steps of -1 do:
smooth all level-k vertices;
transfer the residual from the level-k to level-$(k-1)$

V3 solve $LEVEL1$ equations by sparse Gaussian elimination

V4 for $k = LEVEL1 + 1$ to $LVLF$ in steps of 1 do:
transfer the correction from level-$(k-1)$ basis to level-k;
smooth all level-k vertices

V5 obtain updated guess U'

Here $LVLF$ is the highest level of any element on the triangle tree. At level-k, only the vertices whose (unique) level is k are smoothed; this is in contrast to a standard multigrid v-cycle, where at level-k, all the vertices of level less than or equal to k would be smoothed. The uniqueness of vertex levels implies that each vertex is processed on exactly one level, rather than at several levels, as is typically the case in standard multigrid methods.

The smoothing procedure used is symmetric Gauss-Seidel; one smoothing iteration per level is used in steps V2 and V4. The ordering of equations and unknowns on each level other than $LEVEL1$ is done using a variant of the Reverse-Cuthill-McKee algorithm [21]. The coarse-to-fine and fine-to-coarse grid transfer operations are the standard ones for piecewise linear finite elements, and are essentially defined by a change of basis (from the standard nodal basis to the hierarchical basis). The $LEVEL1$ mesh is ordered using the minimum degree algorithm [25], [21], and the $LEVEL1$

matrix is factored using the general sparse elimination algorithm given in [16], [14].

Procedure V-Cycle is used as a preconditioner for an overall iterative process; there are two basic preconditioned iterative procedures available, specified through the parameter $ISPD$, as summarized in table 2.7. The maximum number of iterations to be used per solution is specified by the parameter $MAXR$. Note as many as $MAXR$ iterations are used each time a system of linear equations is solved, and each Newton iteration may involve the solution of several systems. It should be noted that there are conservative convergence criteria built in to $PLTMG$ to avoid unnecessary computation whenever $ITMAX$ or $MAXR$ is excessively large.

$ISPD$	matrix storage and iteration options
0	nonsymmetric storage, biconjugate gradient iteration
1	symmetric storage, conjugate gradient iteration

Table 2.7.
The values of $ISPD$.

The biconjugate gradient algorithm is fairly robust, but the convergence may be slow if there is poor approximation of the continuous equation on the $LEVEL1$ mesh. One can improve convergence by specifying a larger value for $LEVEL1$, but this forces the solution of a larger set of equations by sparse Gaussian elimination.

The conjugate gradient method is restricted to problems in which the matrix G_u is symmetric, although it may be indefinite. As with the biconjugate gradient method, convergence may be slow if there is poor qualitative resolution of the solution on the coarse ($LEVEL1$) grid (e.g., the coarse grid matrix has fewer negative eigenvalues than the finest grid). This condition can be remedied by choosing a larger value for $LEVEL1$.

2.7. The IP and W Arrays

W is a real array of length $LENW$; all internal storage for $PLTMG$ is allocated from this array. IP is an integer array of length 100 containing various user specified parameters, and internally generated parameters, switches, flags, and pointers.

1	*NT*	26		51	*NTR*	76	
2	*NV*	27	*FILNUM*	52	*NVR*	77	*KSCALE*
3	*NC*	28	*ICRTR*	53	*NCR*	78	*LVLR*
4	*NB*	29	*ICRTW*	54	*NBR*	79	*KFUN*
5	*ISPD*	30	*IUSRSW*	55	*MXT*	80	*NRGN*
6	*NVTRGT*	31	*IFLAG*	56	*MXV*	81	*JTIME*
7	*IPROB*	32	*AFLAG*	57	*MXC*	82	*JHIST*
8	*IADAPT*	33	*YFLAG*	58	*MXB*	83	*JPATH*
9	*LEVEL*1	34	*OFLAG*	59	*MXJB*	84	*JUSR*
10	*IREFN*	35	*CFLAG*	60		85	*IUU*
11	*ITMAX*	36	*IEVALS*	61	*IGRSW*	86	*IUDOT*
12	*MAXR*	37	*ITNUM*	62	*MAG*	87	*IU*0
13	*DIGITS*	38	*LVLF*	63	*IX*	88	*IU*0*DOT*
14	*IDBC*	39	*NTF*	64	*IY*	89	*IUR*
15	*IUNIT*	40	*NVF*	65	*LVLS*	90	*IUL*
16	*LENW*	41	*MAXT*	66	*IFUN*	91	*ITRI*
17		42	*MAXV*	67	*NCON*	92	*IVERT*
18		43	*MAXZ*	68	*ISCALE*	93	*IEE*
19		44	*ITEMP*	69	*NX*	94	*IBB*
20		45	*LENJA*	70	*NY*	95	*IZ*
21		46	*LENA*	71	*NZ*	96	
22		47	*LENJU*	72	*MXCOLR*	97	
23	*IEVSW*	48	*LENU*	73	*IDEVCE*	98	
24	*NP*	49	*STOR*	74	*NUMBRS*	99	
25	*JFUN*	50		75	*LINES*	100	

Table 2.8.
The IP array.

A list of the currently used locations and their names appears in table 2.8. There are nine major groupings of parameters; a brief accounting of each follows. (Not all entries of IP are currently in use.)

$IP(1)-IP(16)$ are control parameters associated with $PLTMG$. Most of these parameters have been defined elsewhere, but are briefly summarized in table 2.9.

I	$IP(I)$	definition
1	NT	number of user triangles
2	NV	number of user vertices
3	NC	number of curved edges
4	NB	number of boundary edges
5	$ISPD$	symmetric/nonsymmetric switch
6	$NVTRGT$	target value for number of vertices
7	$IPROB$	problem type
8	$IADAPT$	refinement switch
9	$LEVEL1$	coarse grid level for multigrid iteration
10	$IREFN$	number of levels of uniform refinement
11	$ITMAX$	maximum number of damped Newton iterations
12	$MAXR$	maximum number of multigrid iterations
13	$DIGITS$	digits of accuracy desired in computed solution
14	$IDBC$	special Dirichlet vertex number
15	$IUNIT$	device number for printed output
16	$LENW$	length of the work array W

Table 2.9.
PLTMG control parameters.

$IDBC$ should normally be set to zero; in the special case when a singular Neumann problem has been specified, so that the solution is computed only up to an arbitrary additive constant, then $IDBC$ can be set to some vertex (between 1 and NV) in order to specify the solution at that point, and thus make it unique. This will be done through the function GXY as with other Dirichlet points, but the special point need not be on the boundary.

The parameters stored in $IP(23)-IP(25)$ are used by $PLTEVL$. They will be discussed in more detail in chapter 7. The parameters stored in $IP(27)-IP(30)$ are used by the driver program $ATEST$.

I	$IP(I)$	definition
36	$IEVALS$	number of function evaluations used on last call
37	$ITNUM$	number of Newton iterations used on last call
38	$LVLF$	maximum level of an element on the triangle tree
39	NTF	total number of triangles on the triangle tree
40	NVF	number of vertices in the current refined mesh
41	$MAXT$	maximum size of triangle tree data structure
42	$MAXV$	maximum number of vertices allowed
43	$MAXZ$	maximum storage for matrices and workspace
44	$ITEMP$	number of temporary arrays of length $MAXV$
45	$LENJA$	integer storage required for fine grid matrix
46	$LENA$	real storage required for fine grid matrix
47	$LENJU$	integer storage required for $LEVEL1$ matrix
48	$LUNU$	real storage required for $LEVEL1$ matrix
49	$STOR$	total storage in W currently used

Table 2.10.
Output data from PLTMG.

They are discussed more fully in chapter 9. $IP(31) - IP(35)$ are flags set by $PLTMG$ and other routines; they are documented in section 2.8 and in chapters 3, 7, and 8.

Most of the parameters $IP(36) - IP(49)$ provide useful information about the internal state of $PLTMG$. Included are many parameters dealing with the partitioning and use of the storage array W. These are summarized in table 2.10.

$PLTMG$ uses the formula

$$LENW = 2 \cdot (NV + NC) + 4 \cdot (NT + NB + 12) \\ +(50 + KP + KS) \cdot MAXV + 640$$

to determine the value of $MAXV$, the number of vertices of the finest mesh that can be accommodated using the available storage in W. The integer $KP = 6$ if $IPROB \neq 8$ and $KP = 0$ if $IPROB = 8$. The integer $KS = 12$ if $ISPD = 0$ and $KS = 0$ if $ISPD = 1$. The basis for this formula is partly empirical, since there is no simple way to estimate the storage necessary for sparse Gaussian elimination of the $LEVEL1$ mesh; indeed, this can change dynamically as the parameter $LEVEL1$ is reset.

$PLTMG$ uses the formula

$$STOR = 2 \cdot (NV + NC) + 5 \cdot NTF + (16 + KP + KS') \cdot NVF$$
$$+ LENJA + LENA + LENJU + LENU + 640$$

to compute the amount of storage actually used. Here the integer $KS' = 4$ if $ISPD = 0$ and $KS' = 0$ if $ISPD = 1$.

$IP(51) - IP(59)$ are input parameters associated with $TRIGEN$; these parameters will be discussed in more detail in chapter 3. The graphics control parameters are stored in $IP(61) - IP(75)$, and will be discussed in more detail in chapters 4, 5, and 6. $IDEVCE$ is a switch available to the test driver $ATEST$ and is not directly used by $TRIPLT$, $GPHPLT$, or $INPLT$. The parameters $IP(77) - IP(80)$ are used by subroutine $SKELTN$. They will be discussed in more detail in chapter 8. Finally, $PLTMG$ uses entries $IP(81) - IP(95)$ as pointers to permanent and temporary arrays within the workspace W.

2.8. Flags and Error Conditions

There are four flags currently set by $PLTMG$. The flags are defined in table 2.11. The interpretation of the return values for these flags is summarized in table 2.12. $PLTMG$ checks the input arrays for obvious errors and reports them through negative values of $IFLAG$. These error conditions are summarized in table 2.13.

I	$IP(I)$	definition
31	$IFLAG$	storage flag
32	$AFLAG$	adaptive refinement flag
33	$YFLAG$	sparse Gaussian elimination flag
35	$CFLAG$	convergence flag

Table 2.11.
Error flags.

2.9. Printed Output

The level-1 continuation and refinement procedures used by $PLTMG$ provide a limited amount of written output. This output is designed to briefly summarize the current state of the computation. More comprehensive output is available from the graphics routines

$IFLAG$	interpretation
0	normal return
16	insufficient storage
41	triangle tree storage exhausted
42	maximum storage for grid functions exceeded
43	maximum storage for matrices exceeded
$AFLAG$	
6	the mesh contains approximately $NVTRGT$ vertices
8	no refinement requested ($IADAPT = 2$ or 4)
13	solution has requested accuracy
16	insufficient storage ($IFLAG$ also set)
$YFLAG$	
0	normal return
$i > 0$	zero pivot ($IFLAG$ also set)
$CFLAG$	
0	normal return
7	maximum failed continuation steps
11	maximum damped Newton iterations
19	sufficient decrease criteria not satisfied

Table 2.12.
PLTMG flag values.

$IFLAG$	interpretation
0	normal return
−100	an illegal node number was found in $IBNDRY$
−110	an illegal midpoint number was found in $IBNDRY$
−120	an illegal edge type was found in $IBNDRY$
−130	a boundary vertex that was not referenced exactly two times in $IBNDRY$ was found
−140	an illegal node number was found in $ITNODE$
−150	a vertex was referenced fewer than three times in $IBNDRY$ and $ITNODE$ combined
−160	a triangle with a boundary edge not specified in $IBNDRY$ was found
−170	the union of triangles did not have NB boundary edges
−180	NV, NC, NT, or NB was not properly defined
−190	two overlapping triangles were found

Table 2.13.
PLTMG data entry errors.

$GPHPLT$ and $TRIPLT$. All formats are designed for output devices having a minimum of 80 characters per line. Output is directed to device $IUNIT$, where $IUNIT = IP(15)$.

A banner is printed specifying the task (level-1 continuation or refinement), and the number of vertices and level number. Additionally, eight columns of numbers are printed. A typical example of such output is illustrated below:

```
                 refinement level = 4    nv =   39

      lambda        rho     lambda dot    rho dot   determinant eigenvalue
0  2  0.000E+00  0.263E+00  0.000E+00  0.000E+00  0.281e  2   0.000E+00
```

The first column contains the current value of $CFLAG = IP(35)$ (in this example, $CFLAG = 0$). The second contains the value of $ITNUM = IP(45)$, the actual number of approximate Newton iterations used. The next four columns contain the current values of the parameter λ, the functional ρ, and their derivatives with respect to arclength along the current solution manifold (in this example, both are zero, since the problem used in this example had no parameter λ). The next column contains the value of the determinant of the factored $LEVEL1$ matrix. The column labeled "eigenvalue" gives an approximation to the smallest singular value of the Jacobian matrix $\mathcal{G}_u$. The singular value (technically always nonnegative) is given a sign chosen such that the inner product of the corresponding singular vectors is positive. This forces the singular value to change sign at a simple singular point, allowing the use of a bisection/secant algorithm for accurately computing the singular point. The approximate singular value is not computed when $IPROB = 8$.

The approximate left and right singular vectors are generated by an inverse iteration technique using zero as the shift, so these vectors are accurately computed only near singularities. (Fortunately, it is only near singular points where the algorithms we use require accurate singular vectors).

Chapter 3

Subroutine TRIGEN

Subroutine *TRIGEN* can be used to generate the initial triangulation for *PLTMG* (i.e., the arrays *ITNODE*, *IBNDRY*, *VX*, *VY*, *XM*, and *YM*) from information describing the shape of Ω.

In general, for simple domains that can be adequately defined by a few triangles, it is usually easier to enter triangle data rather than the data required by *TRIGEN*. However, if the domain is complicated in shape, has internal interfaces (straight or curved) and/or symmetry axes that the triangulation should respect, then *TRIGEN* might significantly reduce the amount of data that the user must generate and enter.

TRIGEN is called using the statement:

Call TRIGEN(VX, VY, XM, YM, ITNODE, IBNDRY, JB, GRADE, HMAX, IP, W)

The meaning of these parameters is discussed below.

3.1. The Domain and the Boundary

The domain Ω is viewed as the union of NTR simply connected subregions Ω_i, $1 \leq i \leq NTR$. The regions need not be convex, and the case $NTR = 1$ is not excluded. A shared boundary between two subregions (an internal interface) will be respected by the triangulation process; that is, the interface will be represented as one or more triangle edges in the triangulation. Such a union of subregions is called a *skeleton*.

The boundary of each Ω_i should be a simple closed curve that does not intersect itself. Thus, for example, if Ω has a hole, adding a single cut between the outer boundary and the hole will not be adequate. At least two subregions will be required in this case.

Having decomposed the domain into NTR subregions, we decompose the boundaries of the subregions into NBR edges b_i, $1 \leq i \leq NBR$. Each edge has two endpoints ν_i^j, $1 \leq j \leq 2$, and if it is a curved edge, it will have a midpoint μ_i. All curved edges will be approximated by the circular arc passing through the midpoint and endpoints. Globally, the vertices are labeled v_k, $1 \leq k \leq NVR$ and the midpoints are labeled m_k, $1 \leq k \leq NCR$. The intersection of any two edges should be at most one common endpoint.

Edges lying on internal interfaces may be curved. At the end of the triangulation process, such edges will be approximated by one or more straight triangle edges, whose end points lie on the original circular arc. Curved edges lying on the boundary of Ω will be approximated by one or more curved triangle edges, as allowed by $PLTMG$.

As an example, we consider the square region with a hole illustrated in figure 3.1. In this example, we decompose the region into 2 subregions ($NTR = 2$), using 10 vertices ($NVR = 10$), 12 edges ($NBR = 12$), and 4 midpoints ($NCR = 4$) as shown.

Global numbering of the subregions, edges, vertices and midpoints is arbitrary. Each vertex v_k, $1 \leq k \leq NVR$, is defined by its x and y coordinates, specified in the arrays VX and VY, in a fashion analogous to $PLTMG$. The VX and VY arrays for this example are shown in table 3.1.

Each midpoint m_k, $1 \leq l \leq NCR$, is defined by its x and y coordinates, stored in the arrays XM and YM, as in $PLTMG$. The XM and YM arrays for this example are shown in table 3.1.

In a fashion similar to $PLTMG$, an edge is defined by five integer parameters - two (global) vertex numbers, a (global) midpoint number, and an edge type, and a label. The possible edge types are:

- 1: natural boundary conditions
- 0: internal edge
- -1: Dirichlet boundary conditions

On output, the label of a boundary edge will be assigned to all boundary edges of the triangulation that lie on that edge. Labels for internal edges are ignored. Edge data is entered in the $IBNDRY$ array in a fashion exactly analogous to $PLTMG$. In our example, we

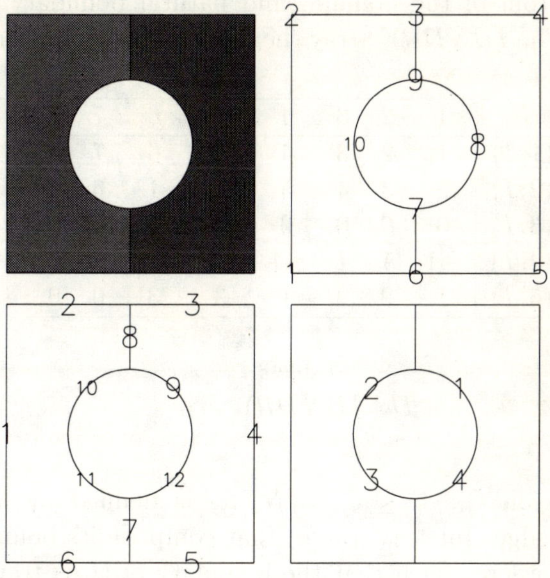

Fig. 3.1. *Clockwise, from upper left: example domain decomposed into two subregions; vertex numbers; midpoint numbers; boundary edge numbers.*

I	$VX(I)$	$VY(I)$	$XM(I)$	$YM(I)$
1	-2	-2	$1/\sqrt{2}$	$1/\sqrt{2}$
2	-2	2	$-1/\sqrt{2}$	$1/\sqrt{2}$
3	0	2	$-1/\sqrt{2}$	$-1/\sqrt{2}$
4	2	2	$1/\sqrt{2}$	$-1/\sqrt{2}$
5	2	-2		
6	0	-2		
7	0	-1		
8	1	0		
9	0	1		
10	-1	0		

Table 3.1.
The VX, VY, XM, and YM arrays. $NVR = 10$ and $NCR = 4$.

will assign Dirichlet boundary conditions to the left and right sides and the bottom of the domain, and natural boundary conditions elsewhere. The $IBNDRY$ array then has the form given in table 3.2.

I	1	2	3	4	5	6	7	8	9	10	11	12
$IBNDRY(1,I)$	1	2	3	4	5	6	7	3	8	10	7	7
$IBNDRY(2,I)$	2	3	4	5	6	1	6	9	9	9	10	8
$IBNDRY(3,I)$	0	0	0	0	0	0	0	0	1	2	3	4
$IBNDRY(4,I)$	−1	1	1	−1	−1	−1	0	0	1	1	1	1
$IBNDRY(5,I)$	1	2	2	1	3	3	0	0	4	4	4	4

Table 3.2.
The $IBNDRY$ array.

A subregion Ω_i, $1 \leq i \leq NTR$, is defined by an ordered sequence of edges (at least three) that comprise its boundary. The sequence is ordered such that the boundary of Ω_i is traversed in a counter clockwise direction (thus providing $TRIGEN$ with notions of "inside" and "outside"). Each edge in the sequence shares exactly one endpoint with the edge that precedes it and the edge that follows it in the sequence; the first and last edges in the sequence also share one endpoint. A particular edge can appear only once in the sequence. Unless $ITNODE(3,I) \neq 0$ (see below) the first edge in the sequence is arbitrary.

The arrays $ITNODE$ and JB are used to define the subregions. Column I of $ITNODE$ corresponds to the region Ω_I. The first entries are pointers into the JB array. In particular, locations $ITNODE(1,I)$ to $ITNODE(2,I)$ of the JB array contain the ordered sequence of edge numbers that define Ω_I. The sequence should conform to the standards described above.

$ITNODE(3,I)$ is used to specify certain symmetries the user may wish to impose on the triangulation. Two subregions are congruent if one can be mapped into the other using an affine transformation consisting of a translation, an (orthogonal) rotation, and perhaps simple reflection. If this mapping also induces one-to-one correspondences between the edges, vertices, and midpoints used to define the regions, then the user can request that $TRIGEN$ triangulate the two regions in a similar fashion.

$ITNODE(3,I) = 0$ specifies that Ω_I can be triangulated independently of other regions. $ITNODE(3, I) = J$, $0 < J < I$, specifies that Ω_I can be mapped onto Ω_J using just a translation and rotation. The first entry in the sequence of edges defining Ω_I should be mapped to the first entry of the sequence of edges defining Ω_J. This insures a one-to-one correspondence for all entries in both ordered sequences. $ITNODE(3, I) = -J$, $0 < J < I$, specifies that Ω_I can be mapped onto Ω_J using a translation, rotation, and a reflection. The first entry of the ordered sequence of edges defining Ω_I should correspond to the *last* entry in the sequence of edges for Ω_J. If the sequence of edges for Ω_J is reversed, (giving a clockwise transversal of the boundary) there will be a one-to-one correspondence between entries of the edge sequences for Ω_I with the inverted order edge sequence for Ω_J. If $ITNODE(3, I) \neq 0$, $TRIGEN$ will map the triangulation generated for Ω_J onto Ω_I, insuring the desired symmetry properties of the overall triangulation. Note that this is not a symmetric relation; $ITNODE(3, I) = J$ does not mean $ITNODE(3, J) = I$. In particular, if $| ITNODE(3, I) | \geq I$, TRIGEN will return in an error condition.

I	1	2
$ITNODE(1, I)$	1	8
$ITNODE(2, I)$	7	14
$ITNODE(3, I)$	0	1
$ITNODE(4, I)$	1	2

I	1	2	3	4	5	6	7	8	9	10	11	12	13	14
$JB(I)$	6	7	11	10	8	2	1	3	8	9	12	7	5	4

Table 3.3.
The ITNODE and JB arrays for mapping by rotation.

In our example, Ω_2 can be mapped to Ω_1 either by rotation or reflection. We can insure the triangulation for Ω_2 will be similar to that for Ω_1, either under rotation or reflection. The overall triangulations may be different in the two cases. (We could insure greater symmetry in the triangulation by decomposing Ω into 4 or 8 similar regions instead of 2 and then setting $ITNODE(3, I)$ appropriately.) The $ITNODE$ and JB arrays for the two situations are illustrated in tables 3.3-3.4. $ITNODE(4, I)$ is a label for the region; all the triangles created in Ω_I will inherit its label.

I	1	2
$ITNODE(1, I)$	1	8
$ITNODE(2, I)$	7	14
$ITNODE(3, I)$	0	−1
$ITNODE(4, I)$	1	2

I	1	2	3	4	5	6	7	8	9	10	11	12	13	14
$JB(I)$	6	7	11	10	8	2	1	4	3	8	9	12	7	5

Table 3.4.
The ITNODE and JB arrays for mapping by reflection.

The arrays VX, VY, XM, YM, $ITNODE$, and $IBNDRY$ are overwritten during the triangulation process by the corresponding arrays required as input to $PLTMG$.

3.2. Some Algorithmic Details

$TRIGEN$ triangulates the regions Ω_i sequentially, taking into account shared internal boundaries and the symmetry requirements.

Let t be a triangle with area a and side lengths h_1, h_2, and h_3. The quality of t, $q(t)$, is measured using the formula

$$q(t) = 4\sqrt{3}a/(h_1^2 + h_2^2 + h_3^2) \tag{3.1}$$

The function $q(t)$ is normalized to be one for an equilateral triangle and to approach zero for triangles with small angles. In attempting to compute a high quality triangulation, $TRIGEN$ uses

$$q(t) \geq .6 \tag{3.2}$$

as a test for acceptability of a triangle (sufficiently small interior angles on the boundaries of the Ω_i could cause (3.2) to be violated).

The triangulation process for those regions for which $ITNODE(3, I) \neq 0$ is simple and is carried out by generating the appropriate affine mapping. The triangulation process for subregions with $ITNODE(3, I) = 0$ is somewhat complicated, but embodies three straightforward heuristics.

Given a subregion viewed as a polygon (possibly with curved edges, and interior angles of size π or greater), $TRIGEN$ first tries to reduce the order of the polygon by one by "chopping" off a triangle from the vertex with smallest interior angle. Inequality (3.2), and several less obvious conditions must be satisfied for a successful chop. When the chopping strategy is no longer successful, $TRIGEN$ checks to see if the remaining polygon is convex with eight or fewer sides. If it is, $TRIGEN$ tries to triangulate the entire remaining subregion by adding the centroid as a vertex and connecting it to each boundary vertex. All the resulting triangles must satisfy (3.2) and some other conditions for this strategy to be successful.

If the second strategy fails or is inapplicable, $TRIGEN$ tries to break the polygon into two smaller polygons by connecting two nonadjacent vertices by a straight line. $TRIGEN$ may exclude many potential cuts as geometrically not feasible or otherwise undesirable. From the remaining possibilities $TRIGEN$ picks the cut that maximizes the minimum of the four interior angles the cut creates. $TRIGEN$ then applies the three strategies to the two newly created polygons in recursive fashion.

If all three strategies fail, $TRIGEN$ recalls the possibility that is least undesirable, and does that. After the region has been successfully triangulated, $TRIGEN$ tries to improve the triangulation by (locally) rearranging edges and adjusting vertex locations such that the criterion (3.2) is optimized.

The user can control the triangulation process to some extent through the parameters $HMAX$ and $GRADE$. Element size is controlled by $HMAX$, the largest triangle edge length (approximately). If $HMAX \leq 0$, $TRIGEN$ will make the triangles as big as possible, consistent with its other objectives. Setting $HMAX$ only places an upper bound on triangle sizes; the sizes of the triangles actually generated depend strongly on the geometry of the Ω_i and may not achieve the bound.

$GRADE$ is (approximately) the largest ratio of sizes of elements sharing a common edge ($1/GRADE$ is the smallest ratio). $GRADE$ should be set on the interval $1.5 \leq GRADE \leq 2.5$; values outside this interval are set to the appropriate end point. Generally speaking, smaller values of $GRADE$ result in smoother transitions from regions of large elements to those of smaller elements, and a higher overall quality measured by (3.2). On the other hand, larger values of $GRADE$ tend to produce meshes with fewer elements, more rapid transitions in element size, and lower overall quality. One may have to experiment to achieve the proper balance between these conflicting objectives.

3.3. The IP and W Arrays

TRIGEN uses the entries in $IP(16)$ and $IP(51)$-$IP(59)$ as input. It writes in locations $IP(1)$-$IP(4)$ and $IP(34)$. The array W is used as workspace. The entries read by *TRIGEN* are summarized in table 3.5; the entries written by *TRIGEN* are summarized in table 3.6.

I	$IP(I)$	definition
16	$LENW$	length of the work array W
51	NTR	number of regions
52	NVR	number of vertices
53	NCR	number of curved edges
54	NBR	number of edges
55	MXT	length of $ITNODE$ array
56	MXV	length of the VX and VY arrays
57	MXC	length of the XM and YM arrays
58	MXB	length of $IBNDRY$ array
59	$MXJB$	length of JB array

Table 3.5.
Entries read by TRIGEN.

I	$IP(I)$	definition
1	NT	number of triangles
2	NV	number of vertices
3	NC	number of curved edges
4	NB	number of boundary edges
34	$OFLAG$	error flag

Table 3.6.
Entries written by TRIGEN.

TRIGEN requires

$$LENW \geq 15 + 6 \cdot MXV + 6 \cdot NTR + 6 \cdot NBR + 5 \cdot NCR + LI + LJ$$

LI and LJ are dimensions for several of the internal data structures used by *TRIGEN*. It appears difficult to give theoretical *a priori* bounds for LI and LJ that are of practical interest; however, in most situations taking LI to be 21 and LJ to be $2 \cdot MXV$ seems more than adequate.

$TRIGEN$ sets the error flag $OFLAG$; various error conditions are summarized in table 3.7.

$OFLAG$	interpretation
0	normal return
16	insufficient workspace in W
55	insufficient storage in $ITNODE$
56	insufficient storage in VX and VY
57	insufficient storage in XM and YM
58	insufficient storage in $IBNDRY$
$100 + J$	error in $ITNODE(3, J)$
301	failure of all three heuristics

Table 3.7.
TRIGEN flag values.

Negative values of OFLAG correspond to various data entry errors, and are summarized in table 3.8.

$OFLAG$	interpretation
0	normal return
−100	an illegal node number was found in $IBNDRY$
−110	an illegal midpoint number was found in $IBNDRY$
−120	a region with fewer than 3 edges was found
−130	an illegal edge number was found in JB
−140	an edge was referenced by 0 or more than 2 regions
−150	an inconsistent edge type was found in $IBNDRY$
−160	two consecutive edges in JB do not share an endpoint
−170	an edge occurred twice in an ordered sequence in JB
−180	NVR, NCR, NTR, or NBR was improperly defined
−190	a region was specified in clockwise order

Table 3.8.
TRIGEN data entry errors.

Chapter 4

Subroutine TRIPLT

Subroutine *TRIPLT* is the graphics package associated with *PLTMG*. As with *GPHPLT* and *INPLT*, *TRIPLT* is a self-contained, portable FORTRAN program, addressing the graphics output device through subroutines *PLINE*, *PFILL*, and *PLTUTL*. The specifications for these routines are given in chapter 10.

TRIPLT is called using the statement:

Call TRIPLT(TITLE, QXY, IP, W)

In the example plots for this chapter, we solved Laplace's equation in a circle of radius one with a crack along the positive x axis. Homogeneous Dirichlet boundary conditions were imposed above the crack and homogeneous Neumann boundary conditions were imposed below. Non-homogeneous Dirichlet boundary conditions were imposed on the circular boundary such that the true solution is $r^{1/4}\sin(\theta/4)$, the leading term in the singularity due to the crack tip. The output device was a PostScript laser printer; we emulated the color features of *TRIPLT* using gray scale.

4.1. Parameter Definitions

TRIPLT uses fourteen variables in the *IP* array in addition to those used by *PLTMG*. These are summarized in table 4.1. The most complicated switch for *TRIPLT* is *IFUN*, which specifies the function to be plotted. The various values for this switch are summarized in table 4.2.

The variable *TITLE* is a *character*80* variable; it may contain any (printable) ASCII character string. *TRIPLT* displays the

I	$IP(I)$	definition
34	$OFLAG$	error flag
62	MAG	window magnification factor
63	IX	(IX, IY) are the window center coordinates
64	IY	
65	$LVLS$	refinement level to use for graphics
66	$IFUN$	alternate function switch for $TRIPLT$
67	$NCON$	number of contours
68	$ISCALE$	scale option switch
69	NX	
70	NY	(NX, NY, NZ) is the viewing perspective
71	NZ	
72	$MXCOLR$	maximum number of colors
74	$NUMBRS$	numbering option switch
75	$LINES$	line drawing option switch

Table 4.1.
$TRIPLT$ control parameters.

contents of $TITLE$ as a banner above the graph, allowing one to label a picture.

The function QXY has argument list:

Function QXY(X, Y, U, UX, UY, RL, ITAG, ITYPE)

All arguments are to be considered as input, and (except for $ITYPE$) are exactly analogous to those for the coefficient functions that specify the differential equation in $PLTMG$. If $ITYPE = 1$, $TRIPLT$ is requesting a function value for the scalar function used in the cases $IFUN = \pm 5$. If $ITYPE = 2$, $TRIPLT$ is requesting a function value for the first component of the vector function used in the cases $IFUN = \pm 17$. If $ITYPE = 3$, the second component of the vector function is being requested. $ITYPE = 4$ is used by $SKELTN$.

If $IFUN = \pm 5$ or $IFUN = \pm 17$, $TRIPLT$ evaluates QXY at each vertex of each element in the $LVLS$ mesh. QXY will generally be multi-valued at the vertices because of discontinuities in ∇u_h. For positive values of $IFUN$, $TRIPLT$ computes a weighted average of QXY at each vertex; the weights are proportional to the area of each element containing the vertex. The resulting grid function is then plotted as a continuous piecewise linear polynomial. If $IFUN < 0$, the function is plotted as a discontinuous piecewise linear polynomial.

$IFUN$	displayed function
0	the solution u_h
1	the tangent vector $\dot{u}_h$
2	the right singular vector u_r
3	the left singular vector u_ℓ
4	the Newton residual $\mathcal{G}(u_h)$
±5	the alternate function QXY
±6	the error estimate in $\mathcal{H}^1$ norm
7	elements colored by level
8	elements colored by region
9	elements colored by geometric quality
10	elements colored by maximum angle
11	elements colored by minimum angle
±12	the vector ∇u_h
±13	the vector $\nabla \dot{u}_h$
±14	the vector ∇u_r
±15	the vector ∇u_ℓ
±16	the vector $\nabla \mathcal{G}(u_h)$
±17	the vector function QXY

Table 4.2.
The values of IFUN.

Although there are many possibilities for $IFUN$, they may be generally classified as *surface plots* ($0 \leq |\, IFUN \,| \leq 6$), *triangle plots* ($7 \leq IFUN \leq 11$), or *vector function plots* ($12 \leq |\, IFUN \,| \leq 17$).

$MXCOLR$ is a device dependent constant, stating the maximum number of colors available for use by $TRIPLT$. $TRIPLT$ assumes $2 \leq MXCOLR$. While it is possible to make some interesting plots and contour maps with $TRIPLT$ using only monochrome devices ($MXCOLR = 2$), $TRIPLT$ makes extensive use of available color facilities in producing (shaded) 3-dimensional surface plots for scalar functions, and vector plots.

4.2. Surface Plots

For the surface plots, $0 \le |IFUN| \le 6$, all the relevant functions are continuous, with the (possible) exceptions of the error, which is piecewise constant on triangles, and QXY. If $IFUN = -6$, the error is plotted as discontinuous piecewise constant; if $IFUN = 6$, a continuous piecewise linear grid function is constructed by a weighted average of the errors for triangles sharing each vertex; weighting is done by the areas of the elements. Some example plots are shown in figure 4.1.

$NCON$ contours (colors) are used in each plot; if 3-dimensional surface plots are requested, appropriate shading is done, provided $MXCOLR$ is large enough to support the required number of shades. If $NCON > MXCOLR - 2$, some colors are used for more than one contour. The $LVLS$ mesh in the refined element tree, consisting of all elements with level $\ell_I \le LVLS$, is used in making the drawings. Normally, one should set $LVLS \ge LVLF$ ($LVLF = IP(38)$) to achieve the most accurate representation possible with the available data.

Each picture consists of three parts; the larger graph on the left, and a two-part legend on the right. In the upper right appears a scale relating colors to function values; three scales are available using the option switch $ISCALE$ that is described in detail elsewhere. In the lower right is a smaller version of the main graph; this is used with the parameters MAG, IX, and IY, and (NX, NY, NZ). Three line-drawing options using $LINES$ and triangle/vertex numbering options using $NUMBRS$ are also available.

4.3. Triangle Plots

For the triangle plots, $7 \le IFUN \le 11$, $TRIPLT$ will draw the $LVLS$ triangulation of Ω; the elements will be color-coded to depict some feature of the mesh. Some examples are shown in figure 4.2.

As with the surface plots, each picture has three parts, a main graph and two legends. For the main graph, three line-drawing options are available using $LINES$ and triangle/vertex numbering options are available using $NUMBRS$. The lower right legend is a smaller version of the main picture, used with MAG, IX, and IY, and (NX, NY, NZ). The upper right legend differs according to the value of $IFUN$. If $IFUN = 7$, a legend depicting the correspondence between colors and levels is shown. A similar legend

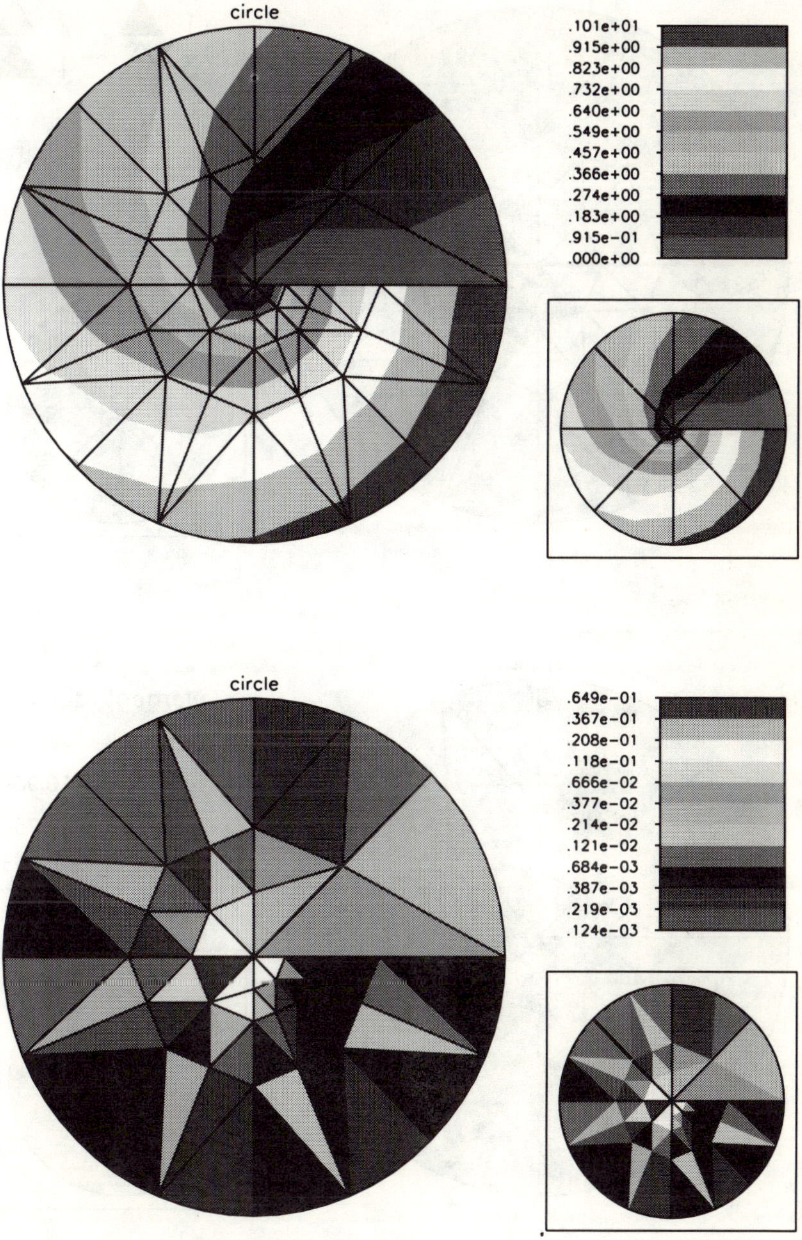

Fig. 4.1. *The solution IGRSW* $= 0$, *top, and the error IGRSW* $= -6$.

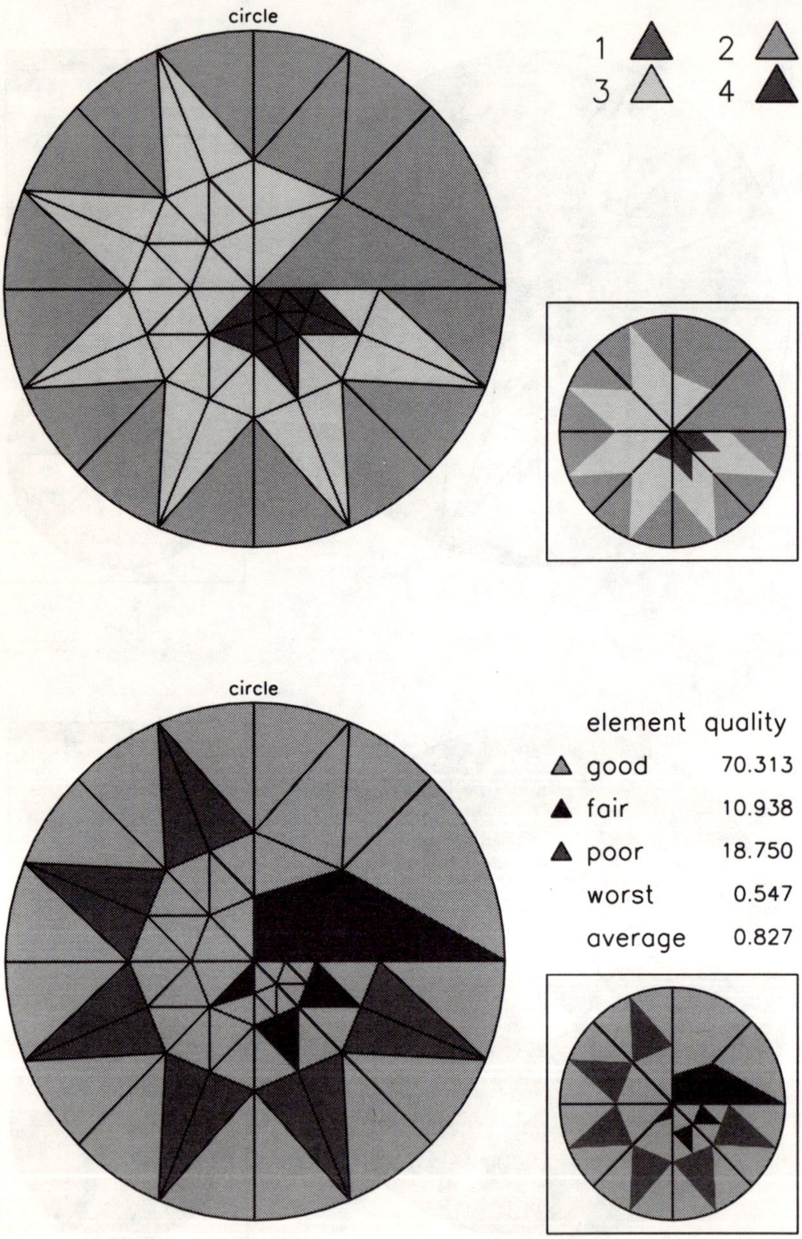

Fig. 4.2. *Triangles colored by level (IGRSW = 7) and by quality (IGRSW = 9).*

is drawn for the case $IFUN = 8$, except that the correspondence is now between regions and colors; two triangles are in different regions if the values of $ITNODE(4,*)$ in the input array to $PLTMG$ are different.

For $9 \leq IGRSW \leq 11$, $TRIPLT$ colors the elements of the triangulation according to their quality, measured by $q(t)$ in (3.1) ($IGRSW = 9$), their largest angle ($IGRSW = 10$), and their smallest angle ($IGRSW = 11$). For each of the three measures, five numbers are printed in the upper right legend. The row labeled "average" refers to the average of that quantity over all elements in the mesh; "worst" reports the smallest value of $q(t)$, largest angle, or smallest angle of all elements. The rows labeled "good", "fair", and "poor" report the percentage of elements in each category and depict the corresponding colors.

For $q(t)$, good means $q(t) \geq \sqrt{3}/2$, fair means $.6 \leq q(t) < \sqrt{3}/2$, and poor means $q(t) < .6$. For large angles, good means $A(t) \leq \pi/2$, fair means $\pi/2 < A(t) \leq 2\pi/3$, and poor means $A(t) > 2\pi/3$ ($A(t)$ is the largest angle). For small angles, good means $\arccos(4/5) \leq a(t)$, fair means $\arccos(13/14) \leq a(t) < \arccos(4/5)$ and poor means $a(t) < \arccos(13/14)$ ($a(t)$ is the smallest angle).

Triangles that are good in terms of $q(t)$ are (necessarily) also good in terms of large and small angles. Those that are fair in terms of $q(t)$ must be good or fair in terms of large and small angles (but not conversely).

4.4. Vector Plots

For the vector plots, $12 \leq | IFUN | \leq 17$, the vector functions in question are discontinuous piecewise constant, except for QXY, which may be developed as continuous or discontinuous piecewise linear. For $IFUN > 0$, the discontinuous function is transformed to a continuous piecewise linear vector function, defining function values at a vertex by weighted averages of the function values in triangles sharing that vertex; weighting is done by the areas of the elements. Some example plots are shown in figure 4.3.

Color plays an important role in the vector plots. Different colors correspond to different directions of flow in the vector field. This is illustrated in the color wheel portion of the upper right hand legend. Different intensities of the same color correspond to the magnitude of the vector; darker shades correspond to smaller magnitudes, and

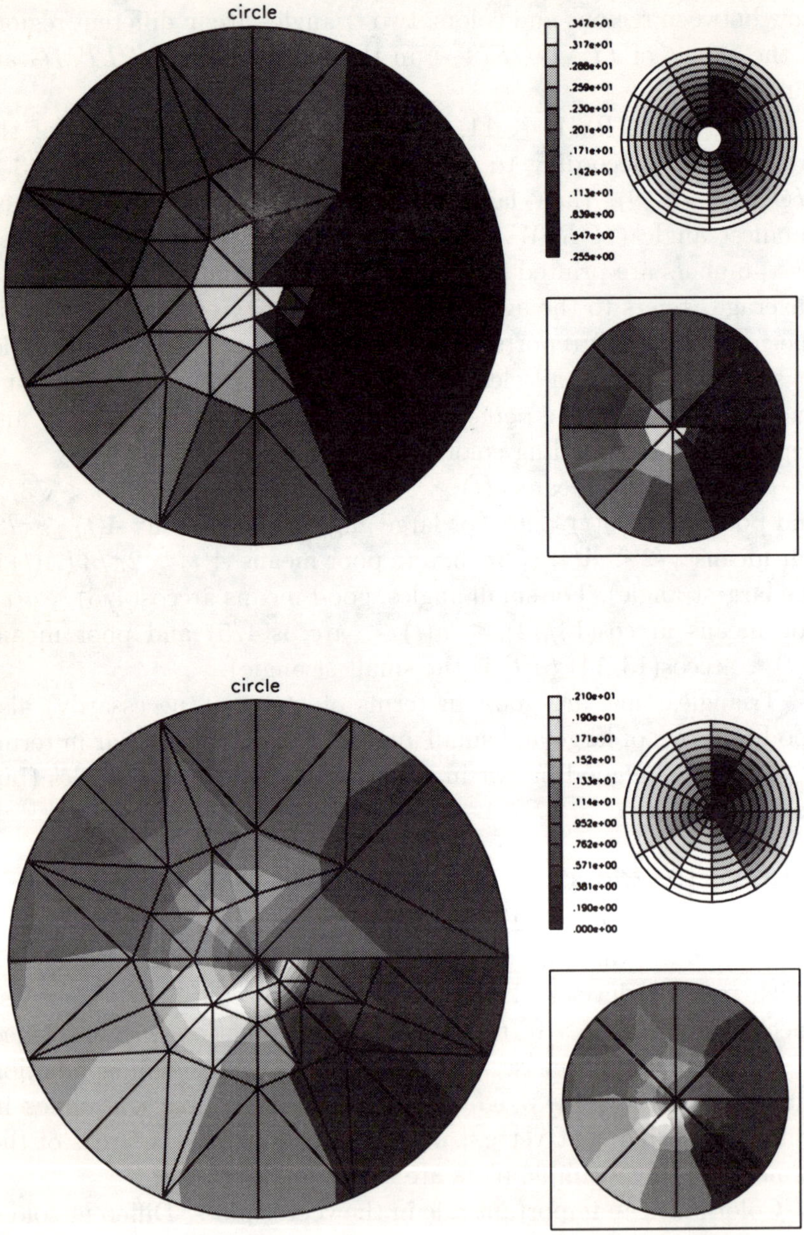

Fig. 4.3. ∇u_h *as a discontinuous function* $(IGRSW = -12)$ *and as a continuous function* $(IGRSW = 12)$.

lighter shades correspond to higher magnitudes. The correspondence between intensity and magnitude is illustrated for an example color in the upper right hand legend. For a vector plot, $NCON$ corresponds to the number of different colors to use for differentiating directions. As with surface plots, three scales for the vector magnitude are available using the option switch $ISCALE$. Two line-drawing options using $LINES$ and triangle/vertex numbering options using $NUMBRS$ are also available.

4.5. The Parameters (NX,NY,NZ)

The triple (NX, NY, NZ) specifies the viewing perspective. It affects all plots, but is intended mainly for the surface plots $0 \leq |IFUN| \leq 6$. For other cases, it is usually best to have $(NX, NY, NZ) = (0, 0, 1)$.

For surface plots, if the direction $d = (NX, NY, NZ)$ is specified, then the three dimensional surface is projected into the plane orthogonal to d, and the required picture is drawn as it would appear to an observer viewing the surface form a line of sight parallel to d. The vectors (NX, NY, NZ) and $(-NX, -NY, -NZ)$ cause the same projection to be computed; however different pictures are generally produced for the two cases. In the former case one observes the projection on the "front" of the plane, and in the latter, one observes the projection on the "back" of the plane. If $MXCOLR$ is sufficiently large, the surface will be shaded relative to a light source directly behind the viewer, imparting some additional 3-dimensional character to the picture.

For triangle and surface plots, the picture is also rotated according to the specified values of (NX, NY, NZ). In this case, however, the drawn surface remains "flat". Some example plots are given in figure 4.4.

The lower right hand legend provides guidance in understanding 3-dimensional surface plots. In this case the legend contains a "flat" version of the main picture, allowing another avenue for orienting oneself with respect to the viewing perspective.

4.6. The Parameters MAG, IX, and IY

The parameters MAG, IX, and IY allow one to focus attention on some particular feature of a picture by expanding the scale and making a window on a small portion of the picture. MAG is the

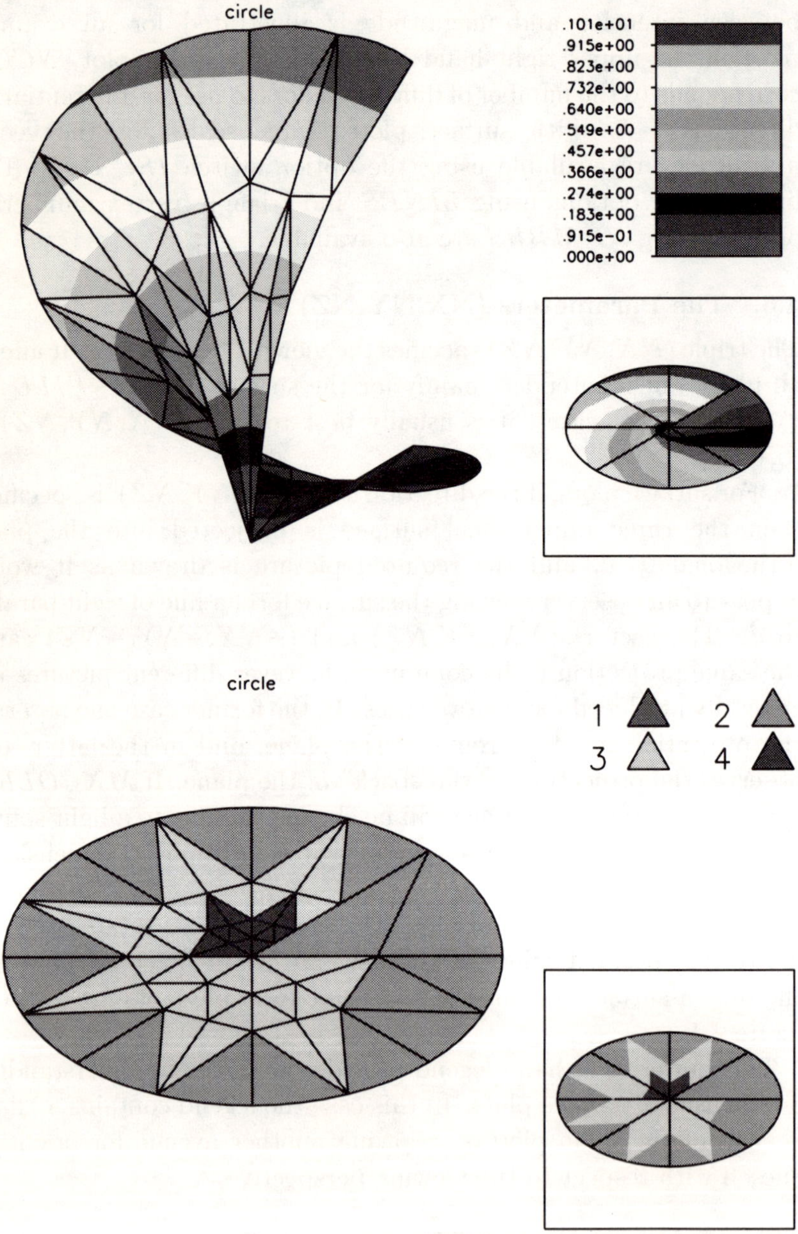

Fig. 4.4. *The case* $IFUN = 0$ *and the case* $IFUN = 7$. *In both pictures* $(NX, NY, NZ) = (1, -1, -1)$.

magnification factor relative to the picture coordinates. For example, if $MAG = 1$ the whole picture will be drawn; if $MAG = 2$, the picture will be scaled by a factor of 2 in both directions, and thus will no longer fit on the output device. One must now choose a window and view only a portion of the picture. If $MAG > 1$ the picture is thought of as having a square $2 \cdot MAG + 1 \times 2 \cdot MAG + 1$ grid imposed in it; This grid will have $2 \cdot MAG + 1$ equally spaced lines in each coordinate direction. The lines are numbered from 0 to $2 \cdot MAG$, left to right, bottom to top. The grid points are doubly indexed in the obvious fashion. With this scaling, exactly four of the square grid blocks, sharing a common vertex, will be visible in the space available on the output device. The common grid point for the four blocks will be the window center for the drawn picture. Thus, if $MAG > 1$, IX and IY should be integers between 0 and $2 \cdot MAG$ specifying the indices for the window center. (If IX or IY is chosen to be 0 or $2 \cdot MAG$, fewer than four grid blocks will be drawn). Some examples are shown in figure 4.5.

As an aid to understanding, the lower right legend contains a copy of the complete picture (corresponding to $MAG = 1$). Whenever $MAG > 1$, a small box is drawn in this legend depicting the portion of the picture appearing in the main graph. For very large magnification factors, the box is replaced by a crosshair locator, since the box would be to small to be visible.

4.7. The Parameters ISCALE, LINES, and NUMBRS

For surface plots and vector plots, $TRIPLT$ allows three possible scalings for the contour lines. These are specified using the parameter $ISCALE$, as summarized in table 4.3.

$ISCALE$	scale
0	linear
1	logarithmic
2	$\sinh^{-1}$

Table 4.3.
The values of $ISCALE$.

For linear scaling, drawn contours are equally spaced with respect to the largest and smallest values of the given function $z(x, y)$. If

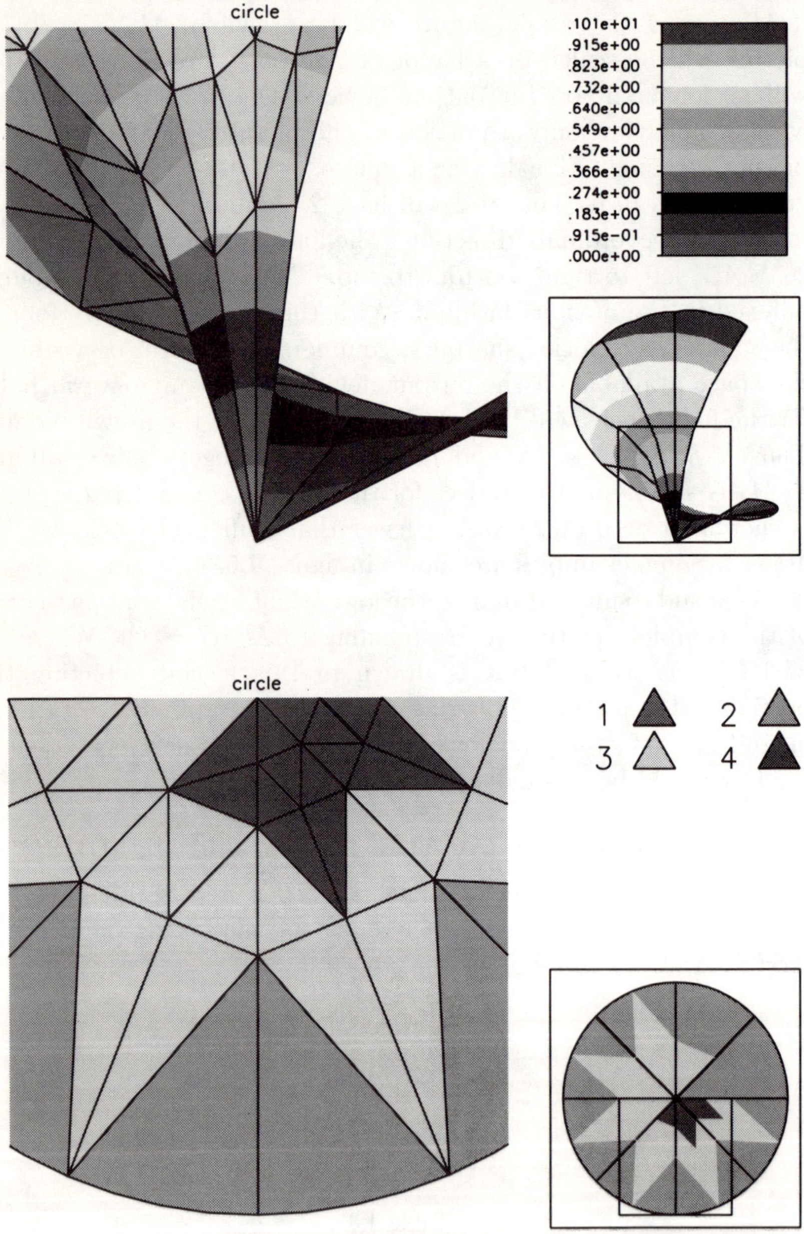

Fig. 4.5. *The case* $IFUN = 0$, $(NX, NY, NZ) = (1, -1, -1)$ *and the case* $IFUN = 7$. *In both pictures* $MAG = 2$ *and* $(IX, IY) = (2, 1)$.

$ISCALE = 1$, then the contours are equally spaced with respect to the largest and smallest values of $\log z$. If $ISCALE = 2$, then the contours are equally spaced with respect to largest and smallest values of the function $\sinh^{-1} z$. The logarithmic scaling clearly requires z to be positive. The $\sinh^{-1}$ scaling is always defined, having a (signed) logarithmic behavior for large $| z |$, and a linear behavior for small $| z |$. If $ISCALE = 1$ and $TRIPLT$ discovers that $z \leq 0$ at some vertex, then it defaults to the $\sinh^{-1}$ scaling. In figure 4.1, the solution u_h was drawn using the linear scale ($ISCALE = 0$), while the error estimate was drawn using the logarithmic scale ($ISCALE = 1$).

Three line drawing options are available in $TRIPLT$; these are specified through the parameter $LINES$, as summarized in table 4.4. If $LINES = 0$, $TRIPLT$ will draw edges of all triangles in the

$LINES$	line drawing option
0	draw triangle edges
1	draw skeleton edges
2	draw contours

Table 4.4.
The values of $LINES$.

mesh. If $LINES = 1$, $TRIPLT$ will draw only boundary edges, and edges separating triangles from different regions (as specified in the $PLTMG$ input $ITNODE(4, *)$).

If $LINES = 2$, and $0 \leq | IFUN | \leq 6$, $TRIPLT$ will draw only boundary triangle edges. Additionally, $TRIPLT$ will draw contour lines separating contours of different colors. Some examples for $LINES = 1$ and $LINES = 2$ are shown in figure 4.6.

The option $LINES = 2$ produces a traditional contour map on monochrome devices, and thus is quite valuable when $MXCOLR = 2$. The $LINES = 1$ option is useful on highly refined meshes, where the $LINES = 0$ option tends to produce unresolved black spots in regions with many small elements.

If $LINES = 2$, and $7 \leq IFUN \leq 11$, $TRIPLT$ will draw boundary triangle edges. Additionally, $TRIPLT$ will draw triangle edges that separate triangles of different colors. This can provide some useful information in the case $MXCOLR = 2$. The option

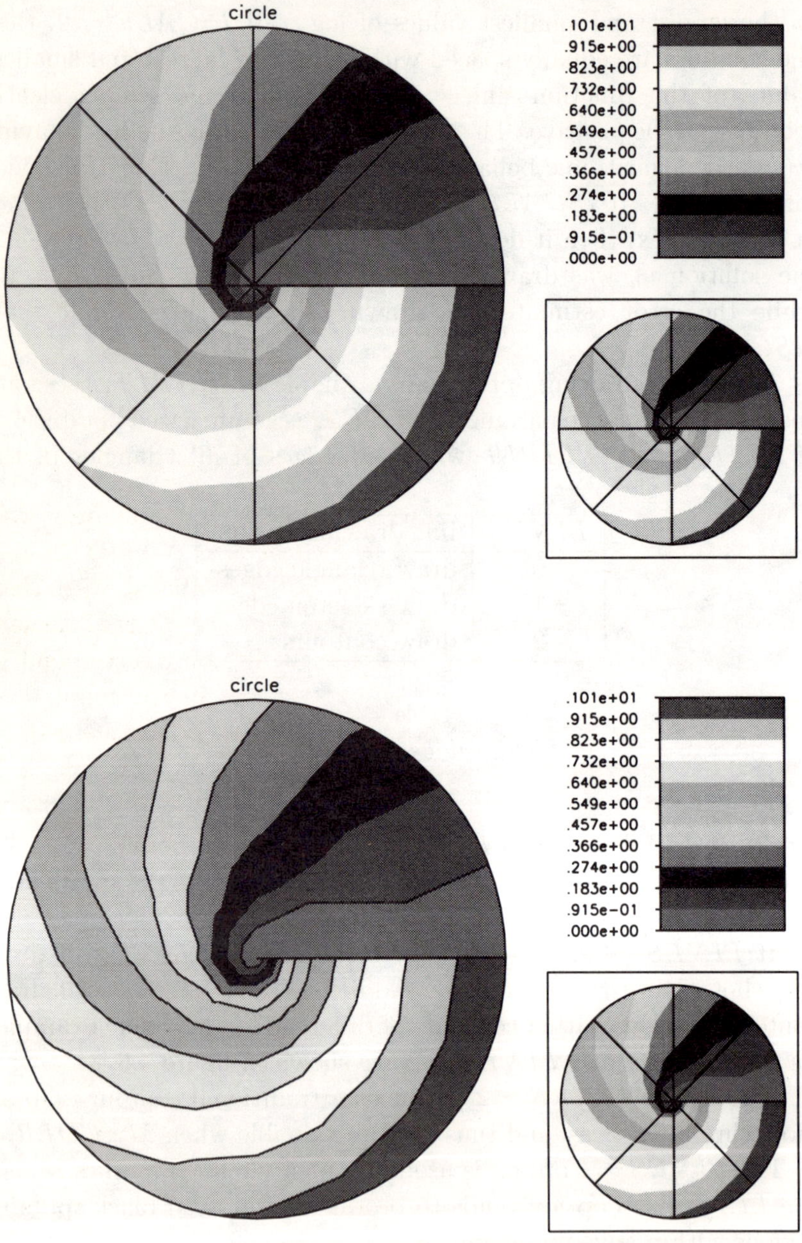

Fig. 4.6. *The case* $LINES = 1$ *and the case* $LINES = 2$. *The corresponding picture for* $LINES = 0$ *is in figure 4.1.*

$LINES = 2$ is not implemented for the case of vector plots, $12 \leq | IFUN | \leq 17$.

Three numbering options are available in $TRIPLT$; these are specified through the parameter $NUMBRS$, as summarized in table 4.5. If $NUMBRS = 1$, then the triangles in the main

$NUMBRS$	numbering option
0	no numbering
1	number triangles
2	number vertices

Table 4.5.
The values of $NUMBRS$.

picture will be labeled with their internal element numbers with respect to the refined element tree in $PLTMG$. If $NUMBRS = 2$, then the vertices in the main picture will be labeled with their internal vertex numbers. One should note that the internal element and vertex numbers used by $PLTMG$ usually don't correspond to the numbering given the elements and vertices in the input data structures. Also, when $NUMBRS \neq 0$ and a surface plot is specified, 3-dimensional plotting is disabled. Trying to make a 3-dimensional plot will result in a "flat" (but numbered) surface, similar to the cases of triangle and vector plots. Some examples are shown in figure 4.7.

4.8. Some Algorithmic Details

The main algorithms of interest in $TRIPLT$ are those for hidden line and surface removal. In the general case of a surface plot, one must make comparisons between various triangles to determine whether a given triangle blocks another with respect to the viewer. Since the triangular mesh is generally unstructured, our goal is to organize the data to minimize the number of comparisons between triangles.

Generally, for surface plots in which $(NX, NY, NZ) \neq (0, 0, 1)$, a partial order is constructed in which elements farthest from the viewer are ordered first, and those closest to the viewer are ordered last. The elements are then drawn and colored in order, with the elements closer to the viewer (possibly) overwriting some elements that are farther away. This procedure works well with video

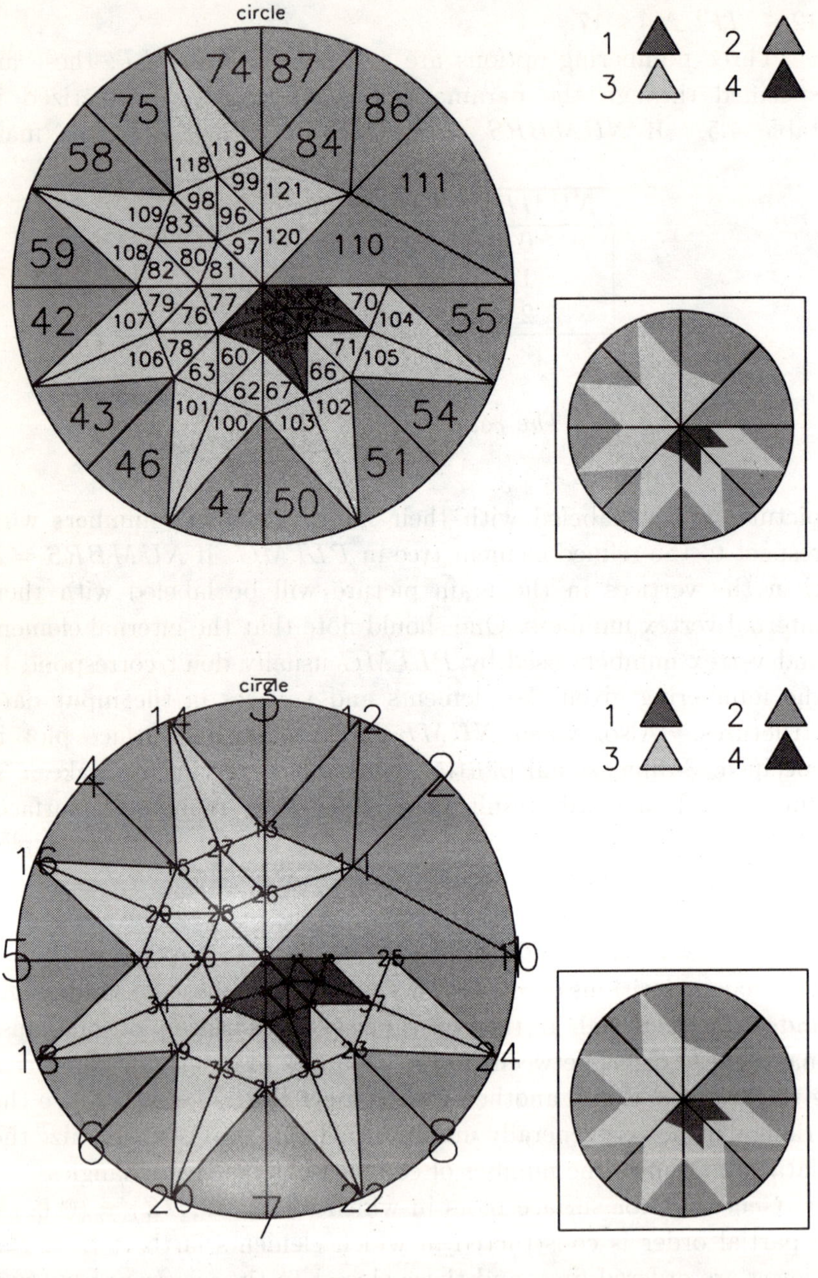

Fig. 4.7. *The case NUMBRS = 1 and the case NUMBRS = 2. The corresponding picture for NUMBRS = 0 is in figure 4.2.*

monitors, and hardcopy devices such as laser printers, but will not work with pen plotters. In the judgment of the author, the gain in speed was well worth the sacrifice in generality (in comparison with older versions of $TRIPLT$, which did explicit hidden line removal).

The notion of distance from the viewer is defined with respect to the x and y coordinates only, so that the same ordering is computed independently of the function being graphed. In any event, a typical element is compared only to elements with which it shares a common edge; it is ordered before any edge neighbors closer to the viewer and after any neighbors farther away. Since any element has at most three neighbors, this greatly limits the number of comparisons necessary, and completely solves the ordering problem for a convex domain with no holes.

Unfortunately, many domains are not convex and have holes, so that elements with boundary edges must be treated as special cases. Thus we make a list of triangles with boundary edges, sort them with respect to the direction (in the (x, y) plane) perpendicular to the (NX, NY) components of the viewing direction. Boundary edges are also sorted by whether they face "backward" or "forward" with respect to (NX, NY). With these preliminary calculations done, all pairs of relevant triangles that *might* conflict are tested and appropriate ordering constraints imposed. On a mesh with NT triangles, the number of boundary triangles is $O(\sqrt{NT})$, so that in the worst case (every boundary element compared with every other boundary element), this will still be only $O(NT)$ work. Since only $O(NT)$ work is required for the interior elements, the overall work is still $O(NT)$.

The error flag $OFLAG$ is set to a nonzero value if there is insufficient storage; its normal return value is zero.

Chapter 5

Subroutine GPHPLT

Subroutine *GPHPLT* provides an assortment of statistical data related to the performance of various algorithms and subroutines in *PLTMG* using a graphical format. It is self-contained, portable FORTRAN; it addresses a graphics output device through subroutines *PLINE*, *PFILL*, and *PLTUTL*, as described in chapter 10.

GPHPLT is called using the statement:

Call GPHPLT(TITLE, IP, W)

In the example output for this chapter, we have solved the Bratu problem

$$-\Delta u - \lambda e^u = 0$$

on the unit square with homogeneous Dirichlet boundary conditions.

5.1. Parameter Definitions

GPHPLT uses two variables in the *IP* array in addition to those used by *PLTMG*; these are shown in table 5.1. *IGRSW* is an integer between 0 and 14; the available options are summarized in table 5.2. For the cases $1 \le IGRSW \le 12$, a single graph is drawn in the unit square. For the cases $IGRSW = 13$ and $IGRSW = 14$, six graphs are displayed in the rectangle $(0, 1.5) \times (0, 1)$; each graph is drawn in a square of size 0.5. The information displayed in the case $IGRSW = 13$ is relevant for all values of *IPROB*, while the information displayed in the case $IGRSW = 14$ is only relevant for problems solved by continuation ($IPROB \neq 8$). The case $IGRSW = 0$ also displays its output in the rectangle $(0, 1.5) \times (0, 1)$.

The displays created by *GPHPLT* use at most eight colors,

I	$IP(I)$	definition
61	$IGRSW$	graph option switch
72	$MXCOLR$	maximum number of colors

Table 5.1.
GPHPLT control parameters.

including foreground and background. However, the display is not significantly impaired if only four or even two colors are available.

The variable $TITLE$ is a *character*80* variable; it may contain any (printable) ASCII character string. $GPHPLT$ displays the contents of $TITLE$ as a banner above the graph.

$IGRSW$	displayed graph
0	the IP array
1	the continuation path
2	timing statistics
3	error estimates for $\mathcal{H}^1$ norm
4	error estimates for $\mathcal{L}^2$ norm
5	error estimates for λ
6	error estimates for ρ
7	multigrid convergence history
8	Newton residual convergence history
9	Newton increment convergence history
10	multigrid convergence for block elimination
11	multigrid convergence for singular value
12	singular value convergence history
13	simultaneously display cases 2, 3, 4, 7, 8, 9
14	simultaneously display cases 1, 5, 6, 10, 11, 12

Table 5.2.
The values of IGRSW.

5.2. Displaying the IP array

If $IPRSW = 0$, $GPHPLT$ displays the entries of the IP array, including both the names of currently used entries and their values. Sample output is shown in figure 5.1 .

square

1	nt	8	26	icvsw	0	51	ntr	0	76		0
2	nv	9	27	filnum	12	52	nvr	0	77	kscale	0
3	nc	0	28	icrtr	5	53	ncr	0	78	lvlr	50
4	nb	8	29	icrtw	6	54	nbr	0	79	kfun	0
5	ispd	0	30	iusrsw	0	55	mxt	4000	80	nrgn	10
6	nvtrgt	400	31	iflag	0	56	mxv	2000	81	jtime	82
7	iprob	0	32	aflag	6	57	mxc	500	82	jhist	122
8	iadapt	3	33	yflag	0	58	mxb	2000	83	jpath	386
9	level1	2	34	oflag	0	59	mxjb	3000	84	jusr	641
10	irefn	2	35	cflag	0	60		0	85	iuu	659
11	itmax	10	36	ievals	4	61	igrsw	0	86	iudot	2853
12	maxr	20	37	itnum	3	62	mag	1	87	iu0	5047
13	digits	6	38	lvlf	5	63	ix	1	88	iu0dot	7241
14	idbc	0	39	ntf	1015	64	iy	1	89	iur	9435
15	iunit	9	40	nvf	397	65	lvls	50	90	iul	11629
16	lenw	150000	41	maxt	5878	66	ifun	0	91	itri	13823
17		0	42	maxv	2194	67	ncon	11	92	ivert	37335
18		0	43	maxz	95818	68	iscale	0	93	iee	46111
19		0	44	itemp	15	69	nx	0	94	ibb	51989
20		0	45	lenja	1886	70	ny	0	95	iz	54183
21		0	46	lena	3374	71	nz	1	96		0
22		0	47	lenju	82	72	mxcolr	248	97		0
23	ievsw	3	48	lenu	210	73	idevce	2	98		0
24	np	1	49	stor	21607	74	numbrs	0	99		0
25	jfun	0	50		0	75	lines	0	100		0

Fig. 5.1. *The IP array.*

5.3. Continuation Path

When $IGRSW = 1$ or $IGRSW = 14$, $GPHPLT$ displays the continuation path generated by the coarse grid continuation procedure. Target points are marked by small boxes, generally using different colors for regular points, limit points, and bifurcation points. All target points (up to fifty) generated by calls to $PLTMG$ since the last call with $IADAPT = 1$ are saved and displayed. Successive points are interpolated using parabolic arcs matching the values of (λ, ρ) and the tangent vectors $(\dot{\lambda}, \dot{\rho})$. Sample output is shown in figure 5.3.

5.4. Timing Statistics

If $IPRSW = 2$ or $IPRSW = 13$, $GPHPLT$ prints a summary of timing statistics for $PLTMG$. Time statistics are given for both the total accumulated time since initialization ($IADAPT = 1$), and for the time spent during the last call to $PLTMG$. The timings are broken down by subroutines that carry out major functions of the package. Depending on the problem, some of these routines may not be called. These subroutines are listed in table 5.3.

A bar graph is drawn illustrating the percentage of time spent in each routine since the last call to $PLTMG$ with $IADAPT = 1$. If $4 \leq MXCOLR$, each bar in the graph is partitioned into a part corresponding to the last call to $PLTMG$ and a part corresponding to all preceding calls. Each partition is then colored with a different color.

subroutine	main function
$TRIANG$	build the triangle tree data structure
$MARKA$	select triangles to adaptively refine and unrefine
$REGRUP$	reformat the triangle tree data structure
$ORDER$	order vertices by level using minimum degree or reverse-Cuthill-McKee algorithms
$SETGR$	compute the graph of sparse matrix on the fine mesh
$PREDCT$	compute the steplength σ for continuation
$LINSYS$	compute the stiffness matrix and right hand side
$SCLASM$	carry out block Gaussian elimination for continuation
$SWBRCH$	switch branches at a bifurcation point
$ENERGY$	compute error estimates for unrefined elements
$RLERR$	compute error estimate for λ
$INTRP$	interpolate solution onto newly refined elements
SNF	compute sparse LDU factors for the $LEVEL1$ matrix
CEV	compute the smallest singular value and vectors
MG	solve linear systems in the block elimination process
*	all other time spent in $PLTMG$

Table 5.3.
Internal subroutines in $PLTMG$.

5.5. Error Estimates

For the cases $3 \leq IADAPT \leq 6$, $GPHPLT$ graphs the function $\log_{10}\mathcal{F}(NV)$, where

$$\mathcal{F}(NV) = \begin{cases} \| \delta u \|_{\mathcal{H}^1(\Omega)} / \| u_h \|_{\mathcal{H}^1(\Omega)} & \text{for } IGRSW = 3 \\ \| \delta u \|_{\mathcal{L}^2(\Omega)} / \| u_h \|_{\mathcal{L}^2(\Omega)} & \text{for } IGRSW = 4 \\ | \delta\lambda |/| \lambda_h | & \text{for } IGRSW = 5 \\ | \delta\rho |/| \rho_h | & \text{for } IGRSW = 6 \end{cases}$$

Here δu is the computed approximation of the error $u - u_h$, $\delta\lambda$ is the computed approximation of the error in λ and $\delta\rho$ is the computed approximation of the error in ρ. While it is hoped that these approximations accurately reflect the state of affairs, the estimates are based on *a posteriori* calculations involving only the computed solution. Some judgment of the validity of such computations may be required.

For the given values of $IGRSW$, the $\log_{10}\mathcal{F}(NV)$ is graphed versus $\log_{10}NV$. All data points (up to the twenty most recent) for which error estimates are available are marked with small (colored) boxes, and joined by linear interpolation. These data points generally correspond to target values $NVTRGT$ used in the adaptive refinement process.

Generally, one expects $\mathcal{F}(NV)$ to behave like

$$\mathcal{F}(NV) \sim \mathcal{C}\,NV^{-q/2} \tag{5.1}$$

Optimally, $q \sim 1$ for the $\mathcal{H}^1$ norm and $q \sim 2$ for the $\mathcal{L}^2$ norm. The data displayed in the graph is fitted by least squares to a function of the form (5.1) and the value of q displayed at the top of the graph as the (estimated) convergence order.

5.6. Convergence Histories

For the cases $7 \le IADAPT \le 12$, $GPHPLT$ graphs the function $\log_{10}\mathcal{R}(k)$, where

$$\mathcal{R}(k) = \begin{cases} \| r_k \|_{\ell_2} \,/\, \| r_0 \|_{\ell_2} & \text{for } IGRSW = 7, 10, 11 \\ \| \mathcal{G}_k \|_{\ell_2} \,/\, \| \mathcal{G}_0 \|_{\ell_2} & \text{for } IGRSW = 8 \\ \| \delta u_k \|_{\mathcal{L}^2(\Omega)} \,/\, \| u_k \|_{\mathcal{L}^2(\Omega)} & \text{for } IGRSW = 9 \\ \| U_k - U_{k-1} \|_{\ell_2} & \text{for } IGRSW = 12 \end{cases}$$

For the cases $IGRSW = 7, 10, 11$, r_k is the residual of a set of linear equations to be solved by the hierarchical basis multigrid method, and k is the iteration (cycle) number. Such linear systems arise in three contexts within $PLTMG$. The matrix involved is always the Jacobian matrix for the discretized nonlinear operator, that has the appearance of a linear elliptic equation; the differences are only in the right hand sides. There is always one system to solve with the Newton residual as right hand side ($IGRSW = 7$). If $IPROB < 8$, then there is a second right hand side associated with the block elimination strategy ($IGRSW = 10$). Finally, there is an inverse iteration system solved to find the smallest singular value of the Jacobian ($IGRSW = 11$). In all three cases, only information about the last twenty cycles of the most recent iteration is saved and displayed.

Information about the convergence of Newton's method is displayed for the cases $IGRSW = 8$ and $IGRSW = 9$. When $IGRSW = 8$, the size of the Newton residual for Newton iteration k relative to the initial Newton residual is displayed. When $IGRSW = 9$ the size of the increment to the solution, relative to the size of the current solution, is plotted. If continuation is used, the estimate also includes the increment $\delta\lambda_k/\lambda_k$. Nominally, on the $LEVEL1$ mesh, the rate of convergence for Newton's method should asymptotically be quadratic; however, on refined meshes, the convergence will usually be linear because the Newton equations are approximately solved using the hierarchical basis multigrid method. As with the other cases, only information about the last twenty Newton iterations in the last call to subroutine $NWTT$ will be displayed.

Finally, if $IGRSW = 12$ the size of the increment to the singular vector for each inverse iteration step k is displayed (since these vectors are always normalized to have unit length in ℓ_2, we need not consider the relative change). If $ISPD = 0$ then the displayed quantity is the larger of the changes in the left and right singular vectors.

In all cases, the function $\log_{10}\mathcal{R}(k)$ is graphed versus the iteration number k. if $\bar{k}$ is the last iteration, then the quantity

$$\mathcal{R}(\bar{k})^{1/\bar{k}}$$

gives an average rate of convergence per step. This number is displayed in a banner above the graph.

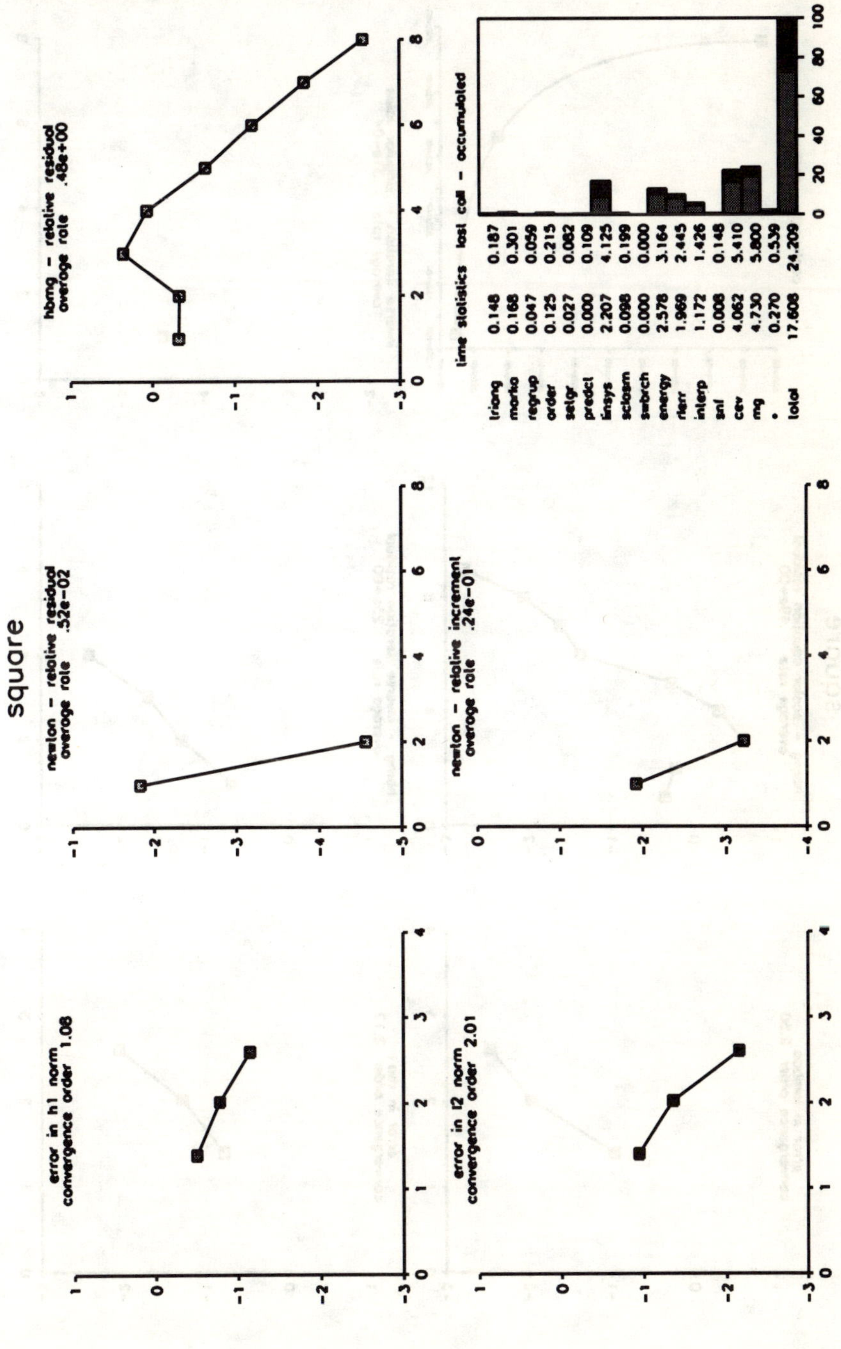

Fig. 5.2. *GPHPLT output for IGRSW* = 13.

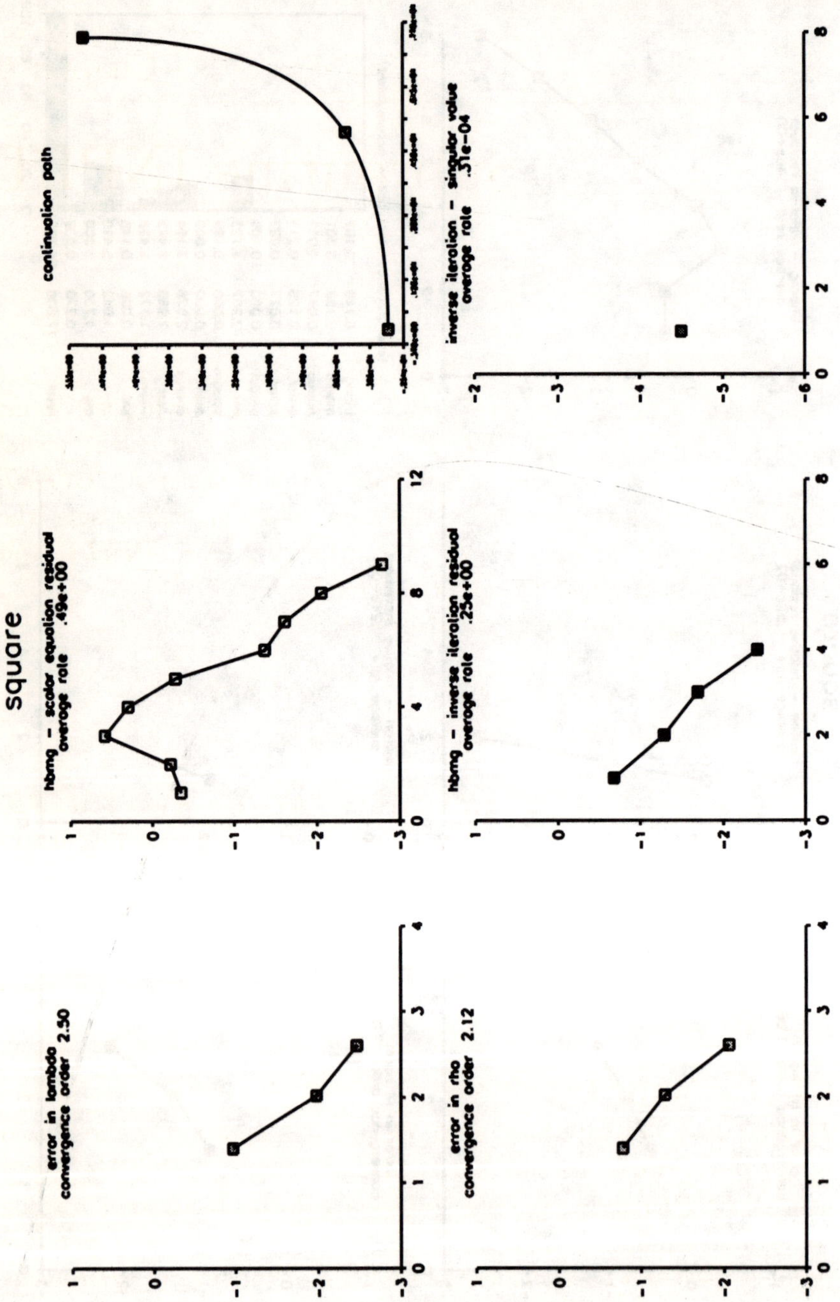

Fig. 5.3. *GPHPLT output for IGRSW* = 14.

Chapter 6

Subroutine INPLT

Subroutine $INPLT$ is a graphics routine for displaying the input data for $PLTMG$ and $TRIPLT$, providing a convenient visual check of the data defining a triangulation or a skeleton. $INPLT$ is a self-contained, portable FORTRAN program, addressing the graphics output device through subroutines $PLINE$, $PFILL$, and $PLTUTL$, as in $TRIPLT$ and $GPHPLT$.

$INPLT$ is called using the statement:

Call INPLT(TITLE, VX, VY, IBNDRY, ITNODE,
XM, YM, JB, IP, W)

The arrays (VX, VY, $IBNDRY$, $ITNODE$, XM, YM, JB, IP, and W) have the same interpretation as the corresponding arrays for $PLTMG$ or $TRIGEN$; $TITLE$ is a *character*80* variable as in $TRIPLT$ and $GPHPLT$. $INPLT$ differentiates $PLTMG$ input data from $TRIGEN$ input data by the value of $ITNODE(3,1)$; it must be zero for valid $TRIGEN$ input and nonzero for valid $PLTMG$ input.

6.1. Parameter Definitions

$INPLT$ uses five variables in the IP array in addition to those used by $PLTMG$; these are summarized in table 6.1. If $PLTMG$ input is used, $INPLT$ draws the triangulation, coloring triangles according to region, similar to $TRIPLT$ with $IFUN = 8$. If $TRIGEN$ input is used, $INPLT$ draws the skeleton of the region, again using different colors for different regions. Some sample output is shown in figure 6.1.

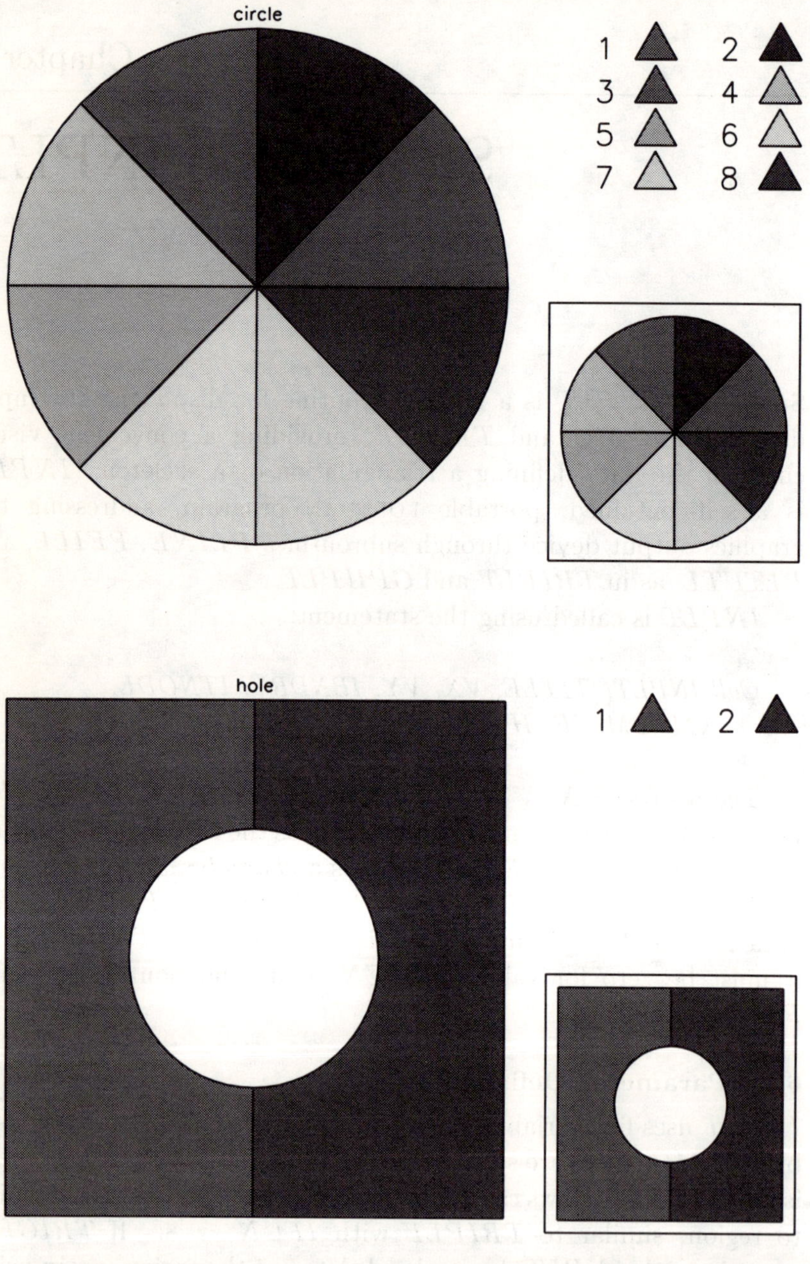

Fig. 6.1. *PLTMG input for the circle with crack; TRIGEN input for the square with a hole.*

I	$IP(I)$	definition
62	MAG	window magnification factor
63	IX	(IX, IY) are the window center coordinates
64	IY	
72	$MXCOLR$	maximum number of colors
74	$NUMBRS$	numbering option switch

Table 6.1.
$INPLT$ control parameters.

The meanings and use of MAG, IX, IY, and $MXCOLR$ are identical to $TRIPLT$. The numbering options that may be specified in $NUMBRS$ are summarized in table 6.2.

$NUMBRS$	numbering option
0	no numbering
1	number triangles
2	number vertices
3	number edges
4	number midpoints

Table 6.2.
The values of $NUMBRS$.

The case $NUMBRS = 1$ is implemented only for $PLTMG$ input; no numbering will occur in this case for $TRIGEN$ input. The numbering used by $INPLT$ always corresponds to the actual numbering used in the data (in contrast to the internal numberings displayed by $TRIPLT$). $INPLT$ was used with the various $NUMBRS$ options to produce figures 2.1 and 3.1, although for those figures the legends on the right hand sides of the pictures were deleted.

Chapter 7

Subroutine PLTEVL

Subroutine $PLTEVL$ evaluates the solution and/or its gradient at a list of user specified evaluation points. $PLTEVL$ is called using the statement:

Call PLTEVL(X, Y, U, UX, UY, IP, W)

7.1. Parameter Definitions

$PLTEVL$ uses four variables in the IP array in addition to those used by $PLTMG$; these are summarized in table 7.1.

I	$IP(I)$	definition
23	$IEVSW$	output option for $PLTEVL$
24	NP	number of evaluation points
25	$JFUN$	alternate function switch
34	$OFLAG$	error flag

Table 7.1.
$PLTEVL$ control parameters.

$IEVSW$ is a switch for controlling the output from $PLTEVL$; the options are summarized in table 7.2. If $IEVSW > 0$, then $PLTEVL$ will carry out several relatively expensive initialization computations; if $IEVSW < 0$, then these computations are skipped. If one makes many calls to $PLTEVL$, without any intervening calls to $PLTMG$ or other output routines, one should set $IEVSW$ to be negative after the first call to avoid repeating the initialization.

$IEVSW$	$PLTEVL$ option
±1	evaluate only u_h
±2	evaluate only ∇u_h
±3	evaluate u_h and ∇u_h

Table 7.2.
The values of $IEVSW$.

The arrays X and Y are of length NP, with $(X(I), Y(I))$ being the I-th evaluation point. The array U is of length NP, with $U(I)$ being the value of the function at the I-th evaluation point. If $IEVSW = \pm2$, the array U can be dimensioned 1 in the calling program to save storage.

The arrays UX and UY are of length NP; their I-th components contain the gradient at the I-th evaluation point. Since the gradient is piecewise constant, it is not uniquely defined along internal triangle edges and at vertices. At such evaluation points a representative (arbitrary) assignment is made from among the possibilities. If $IEVSW = \pm1$, UX and UY can be dimensioned 1 in the calling program.

If a given evaluation point is outside the domain of the problem, the corresponding function and gradient values are set to the minimum value of the function in Ω. For purposes of determining if a point is inside the domain, the circular arc approximation for curved edges is used.

The alternate function switch $JFUN$ is analogous to the switch $IFUN$ for $TRIPLT$, although there are fewer options. The various setting of this switch are summarized in table 7.3.

$JFUN$	alternate function
0	the solution u_h
1	the tangent vector $\dot{u}_h$
2	the right singular vector u_r
3	the left singular vector u_ℓ

Table 7.3.
The values of $JFUN$.

The output error flag, $IP(34)$ is set to -1 if $NP < 1$, $IEVSW$

is out of bounds, or insufficient storage exists for the initialization phase.

7.2. Some Algorithmic Details

The main problem in evaluating a grid function at an arbitrary point (x, y) is determining which element contains the point. Since the meshes in $PLTMG$ are generally nonstructured and nonuniform, this requires searching and testing lists of elements. $PLTEVL$ has an expensive initialization phase, in which elements are sorted by location to limit the lengths of the lists that must be searched for each evaluation point.

This is done by assigning each triangle to a node in a binary tree. We begin by embedding the entire fine mesh in a rectangle, that becomes the root node of the tree. The root rectangle is then bisected, either horizontally or vertically, by connecting a pair of opposing midpoints. This bisection splits the list of triangles into three groups; those completely in the left (top) rectangle, those completely in the right (bottom) rectangle, and a (hopefully small) third group that have nontrivial intersections with both rectangles. The decision whether to divide horizontally, vertically, or not at all, depends mainly on the size of this last group relative to the other two. In any event, if a refinement is made, the two new leaves inherit the lists of elements completely contained in their corresponding rectangles, and the third group of elements remains associated with the father element.

The leaves (son rectangles) then become candidates for further bisection. The overall process creates the binary tree, in which each node is a rectangle, and associated with each node is a short list of triangles. If $IEVSW > 0$, the initialization is carried out and the binary tree created; if $IEVSW < 0$, $PLTEVL$ assumes that the binary tree exists in the workspace W and simply uses the tree to make the function evaluations. If there is insufficient storage to build the binary tree, $PLTEVL$ will return with $OFLAG = -1$; a normal return will have $OFLAG = 0$.

The point $(x, y) \in t$, where t is an element of the triangulation, if and only if all its barycentric coordinates with respect to t are nonnegative (this test is modified slightly for a triangle with a curved boundary edge). The evaluation of the barycentric coordinates requires the assembly and solution of a 3×3 set of linear equations.

The evaluation of a function at the point (x, y) uses two different strategies. In the first, we find a triangle t associated with the leaf of the tree whose rectangle contains the point (x, y); this is done by following a path in the binary tree from the root to the desired leaf. It is relatively inexpensive to determine which of two son rectangles contains the point (x, y).

We evaluate the barycentric coordinates of (x, y) with respect to t; if all are nonnegative, we are done. If one (or two) coordinates are negative, we locate the neighbor element to t corresponding to a negative coordinate; this element is closer to (x, y) than t itself. We then replace t by its neighbor and repeat the test on the new element. In this way we map out a fairly direct path from the seed element to the element that contains the point. Since the seed triangle was associated with the leaf of the binary tree containing the point, we expect the path to contain few elements. In practice, this seems to mean 1–4 elements are tested for each point. In almost all situations, this strategy is successful.

This strategy can fail only if at some step there is no neighbor element, i.e., we arrive at the boundary. If the domain Ω is convex, this means that the point (x, y) is not in Ω. Since we have made no convexity assumption on Ω, it could also mean for example, that we have arrived at a crack and the point is in an element on the other side of the crack. Thus, if the first strategy fails, we build a list of all elements that might contain the point. This is done by marching down the binary tree from the root to the leaf containing the point. The lists of triangles associated with all the nodes along this path are joined to form the list for the given point. This list is checked, beginning with those triangles associated with the leaf, and continuing through the tree towards the root. In this process, we either find an element containing the point, or we exhaust the list and conclude that (x, y) is not in Ω. In practice, this second strategy is required infrequently, even if Ω is not convex.

Chapter 8

Subroutine SKELTN

Subroutine *SKELTN* can be used to generate an input data set for *TRIGEN* from an output triangulation and corresponding solution from *PLTMG*. This data set, called a *skeleton*, can then be used to generate a new triangulation of Ω, providing what amounts to a static rezoning capability for the coarsest mesh.

We envision that *SKELTN* will be most useful in situations where it is important or desirable to have grid lines in the mesh aligned with contour lines of a given function. Generating by hand such a data set might be cumbersome, or even impossible *a priori* if the function in question depended on the solution u. To use *SKELTN*, one can start from *any* description of Ω (either a skeleton for *TRIGEN* or a triangulation for *PLTMG*), and generate a solution of the coarse mesh, or even a refined mesh if necessary. Using this mesh and the corresponding discrete solution, one then calls *SKELTN* to generate a skeleton with subregion boundaries aligned with contours of the discrete solution, or some function of the discrete solution. This skeleton becomes input to *TRIGEN*, which creates a new coarse grid. Since *TRIGEN* preserves subregion boundaries, the resulting triangulation will have element edges corresponding to the original contour lines.

SKELTN is called using the statement:

Call SKELTN(VX, VY, XM, YM, ITNODE, IBNDRY, JB, FRACT, TOL, IP, W, QXY)

The meaning of these parameters is discussed below.

8.1. Input and Output Parameters

The output from $SKELTN$ consists of the arrays VX, VY, XM, YM, $ITNODE$, $IBNDRY$, and JB, that will define a skeleton of the given region.

$SKELTN$ uses the IP array parameters associated with $TRIGEN$ summarized in table 8.1.

I	$IP(I)$	definition
16	$LENW$	length of the work array W
34	$OFLAG$	error flag
51	NTR	number of regions
52	NVR	number of vertices
53	NCR	number of curved edges
54	NBR	number of edges
55	MXT	length of $ITNODE$ array
56	MXV	length of the VX and VY arrays
57	MXC	length of the XM and YM arrays
58	MXB	length of $IBNDRY$ array
59	$MXJB$	length of JB array

Table 8.1.
Entries used by SKELTN.

The values of NTR, NVR, NCR and NBR are written by $SKELTN$, specifying the size of the output data set. The values of MXT, MXV, MXC, MXB, and $MXJB$ are read (but not written), and interpreted by $SKELTN$ as the maximum sizes of the corresponding arrays. $SKELTN$ also sets the output flag $OFLAG = IP(34)$; various error returns are described below. If $OFLAG = 0$, then the data set is valid $TRIGEN$ input, and one may call $TRIGEN$ without further modification of the data.

The input for $SKELTN$ consists of the output from $PLTMG$ in the arrays IP and W. Additionally, there are two real parameters, $FRACT$ and TOL, whose meaning will be more fully described below, and the FORTRAN function QXY. The function QXY has argument list

Function QXY(X, Y, U, UX, UY, RL, ITAG, ITYPE)

All arguments are to be considered as input, and (except for $ITYPE$) are exactly analogous to those for the coefficient functions that specify the differential equation in $PLTMG$. The variable $ITYPE$ takes on integer values between 1 and 4. $1 \leq ITYPE \leq 3$ are used by $TRIPLT$, while $ITYPE = 4$ is reserved by $SKELTN$.

$SKELTN$ uses four variables in the IP array in addition to those used by $PLTMG$; these are summarized in table 8.2.

I	$IP(I)$	definition
77	$KSCALE$	scale option switch
78	$LVLR$	refinement level for contour map
79	$KFUN$	alternate function switch
80	$NRGN$	number of contour lines to use in skeleton

Table 8.2.
$SKELTN$ control parameters.

The $LVLR$ mesh in the refined element tree, consisting of all elements with level $\ell_I \leq LVLR$, will be used by $SKELTN$ in making the skeleton. It is analogous to the parameter $LVLS$ associated with $TRIPLT$. Normally, one should set $LVLR \geq LVLF$ ($LVLF = IP(38)$) to use the most accurate approximate solution. However, if $SKELTN$ should fail, or produce an undesirable skeleton using the $LVLR$ mesh, it is possible that the outcome would be improved by choosing a smaller value for $LVLR$.

The alternate function switch $KFUN$ is analogous to the switch $IFUN$ for $TRIPLT$, although there are fewer options. The various settings of this switch are summarized in table 8.3.

$KFUN$	alternate function
0	the solution u_h
1	the tangent vector $\dot{u}_h$
2	the right singular vector u_r
3	the left singular vector u_ℓ
4	the function QXY

Table 8.3.
The values of $KFUN$.

The function specified by $KFUN$ is the function whose contour lines are to be used in creating the skeleton. If $KFUN = 4$, $SKELTN$ evaluates QXY (with $ITYPE = 4$) at each vertex of each element in the $LVLR$ mesh. QXY will generally be multi-valued at the vertices, because of discontinuities in ∇u_h. Therefore, $SKELTN$ computes a weighted average of QXY at each vertex; the weights are proportional to the area of each element containing the vertex. The resulting grid function is then interpreted as a continuous piecewise linear polynomial.

$SKELTN$ allows the same three scaling options as $TRIPLT$. These are specified using the parameter $KSCALE$, as summarized in table 8.4.

$KSCALE$	scale
0	linear
1	logarithmic
2	$\sinh^{-1}$

Table 8.4.
The values of $KSCALE$.

For linear scaling, contours are equally spaced with respect to the largest and smallest values of the given function $z(x, y)$. If $KSCALE = 1$, then the contours are equally spaced with respect to the largest and smallest values of $\log z$. If $KSCALE = 2$, then the contours are equally spaced with respect to largest and smallest values of the function $\sinh^{-1} z$. As with $TRIPLT$, if $KSCALE = 1$ and $SKELTN$ discovers that $z \leq 0$ at some vertex, then it defaults to the $\sinh^{-1}$ scaling.

Finally, $NRGN$ equally spaced contour lines for the function specified by $KFUN$ will be used as subregion boundaries. The value of $NRGN$ has a significant impact on the resulting triangulation produced by $TRIGEN$. Generally, the larger $NRGN$, the more subregions will be produced. Since the length scales of the subregions are used in determining the length scales of the resulting triangles, triangulating a region with many closely spaced contours will result in many small elements. Using fewer contours will generally result in larger length scales and potentially fewer triangles in the resulting mesh. Thus, choosing the best value for $NRGN$ is often an experimental process.

$SKELTN$ generally should produce a skeleton [1] if the available storage is sufficiently large. $SKELTN$ generates its data structure in the temporary work space portion of the array W. It does not overwrite any of the essential $PLTMG$ data structures. For most of the temporary arrays, the size can be easily estimated *a priori* by $MAXT = IP(41)$ or $MAXV = IP(42)$. However, $SKELTN$ requires one large array, that is allocated all the workspace in W not required for the fixed length arrays. If $SKELTN$ fails because of insufficient storage, the output flag $OFLAG = IP(34)$ is set as in table 8.5.

$OFLAG$	interpretation
0	normal return
41	insufficient storage in array of length $MAXT$
42	insufficient storage in array of length $MAXV$
43	insufficient temporary workspace in W
55	insufficient storage in $ITNODE$
56	insufficient storage in VX and VY
57	insufficient storage in XM and YM
58	insufficient storage in $IBNDRY$
59	insufficient storage in JB

Table 8.5.
$SKELTN$ flag values.

If $41 \leq OFLAG \leq 43$, there is insufficient temporary workspace in W for $SKELTN$ to compute its internal data structures. If $55 \leq OFLAG \leq 59$, then $SKELTN$ has successfully computed the skeleton, but there is insufficient storage of the indicated type for $SKELTN$ to create the corresponding $TRIGEN$ input.

8.2. Some Algorithmic Details

At a conceptual level, the problem of creating a skeleton is similar to the problem of drawing a contour map as in $TRIPLT$. However, in $TRIPLT$, except for the global problem of ordering the triangles

[1] $SKELTN$ is still new enough that it may fail because of bugs. The code includes some numbered stop statements that normally are not executed. These statements are included in an attempt to avoid infinite loops if the internal data structures are corrupted by a bug.

for a surface plot, all the calculations proceed on an element-by-element basis, with the calculation on one element not interacting in any significant algorithmic way with the calculation for any other element. In $SKELTN$ there are significant interactions on a global level, requiring a data structure that can contain the entire contour map.

Thus we develop a data structure in which Ω is partitioned into polygonal subregions. The boundary of a given subregion consists of portions of triangle edges and contour lines. The contours of a piecewise linear polynomial are straight lines in each element, with continuity between elements. Initially, each subregion is contained within a single triangle of the $LVLR$ mesh, and has 3-5 sides, depending on the orientation and number of specified contour lines that appear in the element.

These subregions could, by themselves, be developed into a legal skeleton for $TRIGEN$. However, such a data set would have many more subregions and vertices than necessary. Thus $SKELTN$ performs some transformations on the list of regions, aimed at reducing both the number of subregions, and the number of vertices required to define them.

One basic step is to merge two subregions that share a common boundary into one larger subregion, thus eliminating all the internal edges and vertices along the common boundary. $SKELTN$ attempts to merge smaller subregions to form larger ones, generally respecting the following guidelines:

- Subregions with different labels cannot be merged. The labels are those originally provided by the user in $ITNODE$.

- If the common boundary is a contour edge, then the subregions cannot be merged.

- If the common boundary is not contiguous, then the subregions cannot be merged, as this would create a non-simply connected subregion.

The second guideline may be violated for exceptionally small subregions. These can occur frequently in the original set of subregions, and would result in many small triangles being generated by $TRIGEN$. The user provides guidance to $SKELTN$ through the parameter $FRACT$, which takes on values $0 \leq FRACT \leq 1$.

Generally, if a subregion has an area A satisfying

$$A \leq FRACT \mid \Omega \mid$$

then $SKELTN$ will try to merge it with a larger subregion, even if it must violate the second guideline to do so. Generally, $SKELTN$ tries to create the largest subregions possible within its constraints.

A vertex is said to have *degree k* if it has k incident polygon edges. A *path* is a sequence of connected degree two vertices, generally terminated at each end by a vertex of degree greater than two. $SKELTN$ tries to eliminate unnecessary vertices. To do this, it adheres to the following guidelines:

- A vertex is a candidate for deletion only if it has degree two. This means that the vertex is an internal vertex shared by only two subregions, or a boundary vertex contained in only one subregion.
- A boundary vertex cannot be removed if the two boundary edges it separates have different boundary condition types or different labels. The labels are those originally provided by the user in $IBNDRY$.
- A vertex is removed only if it is (approximately) colinear with the vertices on the path containing the given vertex, or if it is a redundant vertex on a circular arc approximation of the path.

$SKELTN$ examines all degree two vertices, deleting those that are redundant. The user supplied variable TOL is used by $SKELTN$ in numerical tests to determine redundancy. TOL should be regarded as a relative tolerance, as it is independent of the scaling of the (x, y) coordinates. The smaller the value of TOL, the stricter will be the tests for colinearity, etc. Generally, choosing $TOL \sim 10^{-2}$ seems to be a good compromise between having the tolerance too strict, and thus having too many vertices included in the skeleton, and having the tolerance too weak and having insufficient vertices to accurately represent the contour lines.

The data reduction transformations described above maintain a data set corresponding to a valid skeleton. Thus, after the transformations are completed, the remaining subregions and vertices are used to generate the appropriate $TRIGEN$ data structures.

Chapter 9

Program ATEST

Program *ATEST* is the test driver used in the development and testing of subroutines *PLTMG*, *TRIGEN*, *GPHPLT*, *PLTEVL*, *TRIPLT*, *INPLT* and *SKELTN*. *ATEST* is a flexible program in that it accepts simple command strings directing it to call subroutines or perform other tasks. It is not limited to a fixed sequence of tasks on a particular run; any routine can be called as often as desired, with certain parameters reset for each call at the discretion of the user.

ATEST is designed as an interactive program, accepting commands from the user via a CRT. However, it also can be used as a batch program by modifying the machine dependent routine *GETCMD* that is responsible for acquiring the command strings. Details of *GETCMD* are discussed in chapter 10.

9.1. Available Commands

Commands are entered in strings of 80 characters, counting blanks. The syntax of a command can take several forms, but the root command is always a single letter. The commands that are currently recognized by *ATEST* are summarized in table 9.1.

ATEST operates at two levels; *command level*, which is characterized by the prompt "$*$", and *input level*, characterized by the prompts ">" and "?". When *ATEST* is at command level, one can enter one or more command strings.

If one enters a blank line at the command level prompt "$*$", *ATEST* will display a list of the available commands, as illustrated below.

```
commands
```

Command	Action
I	call $INPLT$
T	call $TRIGEN$
S	call $PLTMG$
G	call $GPHPLT$
F	call $TRIPLT$
K	call $SKELTN$
R	reset $TITLE$
D	write data set to a file
L	read data set from a file
U	call $USRCMD$
X	exit

Table 9.1.
Available commands for $ATEST$.

```
t triangulate domain     k create skeleton       d write data file
s solve p. d. e.         f plot function         l read data file
g graph output           i plot input            x exit
r reset title            u call usrcmd
*
```

Input command strings at command level are separated by the symbol "$*$", as shown below.

$$* CMD1 * CMD2 * CMD3 * \ldots * CMDN$$

Here $CMDI$ are individual commands. Commands are processed in the order that they appear in the string. All "$*$" symbols except the first are provided by the user. A syntax error in a given command causes that command, and those that follow it in the current string, to be ignored. After it encounters such an error, $ATEST$ will display the list of available commands, and then simply ask for a new command string by presenting the command prompt "$*$". $ATEST$ will display a statement of each command prior to its execution.

For example, the simplest command strings are just single letters. The command

$$* S$$

would cause $ATEST$ to display the line

```
solve p. d. e.
```

call $PLTMG$, and then return to the user for the next command.

The command line

$$*T*I*S*G*X$$

would cause *ATEST* to first call *TRIGEN* to generate a mesh, then display the resulting mesh using *INPLT*, solve the problem using *PLTMG*, display some output from *GPHPLT*, and exit. Before each command is executed, the appropriate command line is displayed. To keep the driver as simple as possible, *ATEST* does not check the consistency or legality of a command string (e.g., it is up to the user to be sure that he has solved the problem before attempting to graph the solution).

Of course, one may wish to modify parameters between calls to various routines; this is possible using several alternatives. The most simple is to enter input mode for a given command. To do this, simply type the command followed by a period. *ATEST* will respond by listing the integer and real parameters that can be reset for the given command, along with their current values. For example the command

$$*S.$$

would produce the display

```
integer parameters

p  iprob      8        l1 level1     1        m  maxr      20
a  iadapt     1        ir irefn      1        d  digits     6
n  nvtrgt    40        it itmax     10        u  iunit      6
i  ispd       1

real parameters

r  rtrgt   0.000E+00    l  rltrgt  0.000E+00
>
```

These are eleven integer and two real parameters affecting *PLTMG* that can be reset by the user. To the right of each parameter is its current value, while to the left is a one or two letter alias (to avoid having to continually type long names). To reset some variables, at the prompt ">" one enters an input string of the form:

$$> n_1 = v_1,\ n_2 = v_2,\ \ldots\ n_k = v_k$$

Here the n_j refer to variable names or their aliases, and the v_j refer to integer or real values. For example:

$$> IPROB = 1,\ RTRGT = 3.E-1,\ L = 1.,\ IR = 2$$

would reset $IPROB$ to 1, $RTRGT$ to .3, $RLTRGT$ to 1.0, and $IREFN$ to 2. Note the variables are referenced by name or alias; the variable and its new value are separated by an "=" sign. If several variables are to be reset in the same command, different entries are separated by a comma. Integer data must be entered as integers; real data can be entered using integer, fixed point or exponential notation. A syntax error in the input line (e.g., a misspelled variable name) causes the entire command to be ignored, and no variables reset. $ATEST$ will then respond with a "?" prompt, indicating the data should be reentered. After the data has been accepted, $ATEST$ displays the parameter list with the updated values. In our example:

```
integer parameters

p  iprob        1        l1 level1      1        m  maxr        20
a  iadapt       1        ir irefn       2        d  digits       6
n  nvtrgt      40        it itmax      10        u  iunit        6
i  ispd         1

real parameters

r  rtrgt    0.300E+00    l  rltrgt   0.100E+01
>
```

At this point, one may reset additional parameters, or exit input mode by entering a blank line.

One can also reset parameters in command mode (assuming one remembers their names or aliases) by simply typing the reset string after the command. For example

$$* \ S \ IPROB = 1, \ L = 1., \ IR = 2 * F \ IFUN = -6$$

would reset $IPROB$ to 1, $RLTRGT$ to 1.0, and $IREFN$ to 2, and then call $PLTMG$. Then $IFUN$ would be set to -6 and $TRIPLT$ called.

Finally, the G, F, I, D and L commands have a *short form* which allows the resetting of one crucial parameter ($IGRSW$, $IFUN$, $NUMBRS$, $FILNUM$, and $FILNUM$, respectively) without retyping even the alias. For example,

$$* \ F6$$

is the short form for

$$* \ F \ IFUN = 6$$

The short form and the long form of these commands cannot be mixed, e.g.,

$$* \; F6 \; NCON = 10$$

is not valid. To reset more than the crucial parameter, one must use

$$* \; F \; IFUN = 6, \; NCON = 10$$

or enter input mode for the F command.

Most of commands result in a branch to the given subroutine, resetting parameters for that routine, or both. Several commands (D, L, R, U) do not call routines documented elsewhere in this manual, and so are described in more detail below.

9.2. Array Dimensions and Initialization

$ATEST$ has two labeled *common* blocks :

```
common /ATEST1/ IP(100), RP(100)
common /ATEST2/ IU(100), RU(100)
```

The array RP contains floating point parameters used by various routines. Only six entries in RP are currently used; these parameters are defined in table 9.2.

I	$RP(I)$	definition
1	$RTRGT$	target value for ρ in $PLTMG$
2	$RLTRGT$	target value for λ in $PLTMG$
3	$HMAX$	element size bound for $TRIGEN$
4	$GRADE$	grading parameter for $TRIGEN$
5	$FRACT$	small domain tolerance for $SKELTN$
6	TOL	drop tolerance used by $SKELTN$

Table 9.2.
The RP array.

There are five parameters in the IP array that are mainly associated with $ATEST$. These are shown in table 9.3. $FILNUM$ is used by the D and L commands. $ICRTR$ and $ICRTW$ specify the input and output devices used by $ATEST$, and in particular, by the machine dependent routine $GETCMD$ (see chapter 10). $IUSRSW$ is a switch which the user may set within the user supplied routine $USRCMD$. Finally, $IDEVCE$ can be used by the graphics routines $PLTUTL$, $PLINE$, and $PFILL$ for switching between several graphics output devices (see chapter 10).

I	$IP(I)$	definition
27	$FILNUM$	file number for reading and writing files
28	$ICRTR$	command input device number
29	$ICRTW$	command output device number
30	$IUSRSW$	user command switch for branching to $GDATA$
73	$IDEVCE$	graphics output device switch

Table 9.3.
$ATEST$ control parameters.

The arrays IU and RU are not directly used by $ATEST$ or any of the other routines. They are provided to the user for storing integer and real parameters associated with a particular problem. The advantages in using these arrays are that they are saved in the D and L commands, the common block $ATEST2$ can be included in the functions $A1XY$, $A2XY$, etc., where the parameters may be needed, and they can form part of the interface for resetting problem parameters using $USRCMD$.

The user input arrays $ITNODE(4, MXT)$, $VX(MXV)$, $VY(MXV)$, $XM(MXC)$, $YM(MXC)$, $IBNDRY(5, MXB)$, $JB(MXJB)$, and the work array $W(LENW)$ are declared at the beginning of $ATEST$. The sizes of the arrays, MXT, MXV, MXC, MXB, $MXJB$, and $LENW$, are specified at the beginning of $ATEST$ using a *parameter* statement; changing sizes to suit a particular computing environment or problem is thus a simple matter.

To use $ATEST$, the user must provide FORTRAN functions $A1XY$, $A2XY$, FXY, GXY, $P1XY$, $P2XY$, and QXY (QXY may be a dummy function if it is not used). Subroutine $USRCMD$ should be provided, if only as a dummy routine. The user must also supply subroutine $GDATA$, in which the main input arrays for $PLTMG$ or $TRIGEN$, VX, VY, XM, YM, $ITNODE$, $IBNDRY$, and JB, are specified, along with some parameters in IP, RP, and possibly IU and RU.

All the entries of the IP and RP arrays which are not specified by the user through $GDATA$ are given default values in $DATA$ statements at the beginning of $ATEST$. These can be set by the user to give any defaults which are desired, or reinitialized by the user in $GDATA$.

9.3. The D and L commands

The D and L commands are used to save and restore data sets. One can save either the input data or the current solution. Both the D and L commands make use of $IP(27) = FILNUM$. If $0 < FILNUM < 99$, then $TITLE$, the arrays IP, RP, IU, RU, and the portions of W corresponding to the current state of the calculation are written to (D command) or read from (L command) the file specified by $FILNUM$. This file has the name $RW1xx.SV$, where $xx = FILNUM$. For example the commands

$* D\ 12$

(short form) or

$* D\ FILNUM = 12$

(long form) would create the file $RW112.SV$ containing the data described above. All the data is written using unformatted i/o to save space. All i/o is addressed to unit 14.

The command

$* L\ 12$

will restore a data set previously saved in $RW112.SV$. One can use the D and L commands to save the solution at various points along a continuation path, using different values of $FILNUM$ to create different files corresponding to different solutions. One can also save solutions in the current run for postprocessing (graphics, etc.) which can then occur in a later run. We often use the D and L commands as an aid in debugging, allowing us to avoid long complicated command sequences which may be required to generate the state in which the bug occurs; we can generate the state prior to the failure just once, save it, and then restore it as required for debugging. The variable $TITLE$ can be reset (via the R command) to help identify different data sets.

If $0 > FILNUM > -99$, then $TITLE$ and the arrays IP, RP, IU, RU, VX, VY, XM, YM, $IBNDRY$, $ITNODE$, and JB are written to (D command) or read from (L command) the file specified by $FILNUM$. This file has the name $RW0xx.SV$, where $xx = -FILNUM$. The file $RW000.SV$ (which corresponds to $FILNUM = 0$) is reserved by $ATEST$ to save the input data before calling $TRIGEN$, since some input arrays will be overwritten. We have found this option most useful in saving skeletons ($TRIGEN$ input data sets) generated by $SKELTN$. In most other respects, saving input data is similar to saving the solution; unformatted i/o addressed to unit 14 is used, and one can save the input data generated in one run for use in a later run.

9.4. The R command

The command

R

causes $ATEST$ to respond with the prompt "+". At this prompt, the next 80 or fewer characters entered by the user are placed in the *character*80* variable $TITLE$ used by $TRIPLT$, $GPHPLT$, $INPLT$, and the D and L commands. A blank line will result in a blank $TITLE$. Only printable ASCII characters may be used. Subroutine $GETCMD$ receives the character string stored in $TITLE$.

9.5. Subroutine USRCMD

The command

U

causes $ATEST$ to call the user supplied routine $USRCMD$.

Subroutine USRCMD

This routine is written by the user to perform any tasks not covered by other commands. In our experience, the most frequent use of $USRCMD$ has been to reset parameters unique to a particular problem.

$USRCMD$ is affected by the variable $IUSRSW = IP(30)$. If $IUSRSW = 0$, the return from $USRCMD$ will result in the next command in the current command string (if any) to be executed, or $ATEST$ will present the command prompt "*". If $IUSRSW \neq 0$, the return from $USRCMD$ will result in a branch to the user supplied routine $GDATA$ prior to the execution of the next command. This switch is useful if modified parameters affect the geometry of the region, boundary conditions, etc., requiring new definitions for some of the input arrays.

Since the most frequent use of $USRCMD$ is for resetting problem dependent parameters, we now describe how to build an interface within $USRCMD$ allowing one to reset parameters in a fashion similar to input mode for the other commands. The routine that does this is *RESET*, and it is called as follows:

Call RESET(TABLE, ALIAS, INDX, NI, NR,
IU, RU, ICRTR, ICRTW)

$ICRTR$ and $ICRTW$ are device numbers for input and output, and can generally be taken as $IP(28)$ and $IP(29)$, respectively. *IU* and *RU* are integer and real arrays, respectively, containing the integer and real parameters to be reset. It is often convenient to use the *IU* and *RU* arrays provided by $ATEST$ in common block $ATEST2$ for this purpose. $NI \geq 0$ and $NR \geq 0$ are integers specifying the number of integer and real parameters to be reset. The entries in *IU* and *RU* corresponding to these parameters need not be contiguous.

$TABLE$ is a *character*6* array of length $N > NI + NR$, containing the parameter names; if a name is less than six characters, it should be left justified. $RESET$ assumes that the first NI entries in $TABLE$ are names of integer parameters associated with the array *IU*; entries $NI + 1$ to $NI + NR$ are names associated with floating point parameters stored in *RU*. $ALIAS$ is a *character*2* array of length N, containing the aliases for the parameters; if an alias is less than two characters, it should be left justified. $INDX$ is an integer array of length N, containing pointers to the parameter locations in *IU* or *RU*. For $1 \leq I \leq NI$, $IU(INDX(I))$ is the integer parameter with name $TABLE(I)$ and alias $ALIAS(I)$. For $NI+1 \leq I \leq NI+NR$, $RU(INDX(I))$ is the real parameter with name $TABLE(I)$ and alias $ALIAS(I)$.

When $RESET$ is called, it creates a display of the names in $TABLE$, the aliases in $ALIAS$, and the current values of the parameters in *IU* and/or *RU*, and presents the prompt ">" asking for input. Input is supplied as in the input mode format for $ATEST$ described above. The appearance of the "?" prompt indicates an error on the previous input line, and the data should be reentered. Entering a blank line terminates input, and returns from $RESET$ with the updated values contained in the appropriate locations of *IU* and *RU*.

Below we give an example $USRCMD$, associated with the $GDATA$ example from the next section. Here we reset 5 integer and 16 real parameters. The real parameters are coefficients in the functions $A1XY$, $A2XY$, FXY and GXY, allowing the definition and solution of a wide variety of partial differential equations. The first 4 integer parameters in IU are switches allowing the boundary condition type for each of the four sides of the square domain to be independently reset. The fifth integer parameter , $ICONT$, specifies which of the 16 real parameters is to be regarded as the continuation parameter λ. If $ICONT = 0$, then the problem has no continuation parameter ($IPROB = 8$).

```
        subroutine usrcmd
c
            common /atest1/ip(100),rp(100)
            common /atest2/iu(100),ru(100)
            character*6 table(21)
            character*2 alias(21)
            integer indx(21),isv(5)
            save table,alias,indx,ni,nr,isv
c
        data ni,nr/5,16/
        data table( 1),alias( 1),indx( 1)/'left  ','l ',1/
        data table( 2),alias( 2),indx( 2)/'top   ','t ',2/
        data table( 3),alias( 3),indx( 3)/'right ','r ',3/
        data table( 4),alias( 4),indx( 4)/'bottom','b ',4/
        data table( 5),alias( 5),indx( 5)/'icont ','i ',5/
        data table( 6),alias( 6),indx( 6)/'a1x   ','x1',1/
        data table( 7),alias( 7),indx( 7)/'a1y   ','y1',2/
        data table( 8),alias( 8),indx( 8)/'a1u   ','u1',3/
        data table( 9),alias( 9),indx( 9)/'a2x   ','x2',4/
        data table(10),alias(10),indx(10)/'a2y   ','y2',5/
        data table(11),alias(11),indx(11)/'a2u   ','u2',6/
        data table(12),alias(12),indx(12)/'bux   ','bx',7/
        data table(13),alias(13),indx(13)/'buy   ','by',8/
        data table(14),alias(14),indx(14)/'cu0   ','c0',9/
        data table(15),alias(15),indx(15)/'cu1   ','c1',10/
        data table(16),alias(16),indx(16)/'cu2   ','c2',11/
        data table(17),alias(17),indx(17)/'cu3   ','c3',12/
        data table(18),alias(18),indx(18)/'cexp  ','cx',13/
        data table(19),alias(19),indx(19)/'du0   ','d0',14/
        data table(20),alias(20),indx(20)/'du1   ','d1',15/
        data table(21),alias(21),indx(21)/'eu0   ','e0',16/
c
c       save integer parameters
c
        do 1 i=1,5
            isv(i)=iu(i)
    1   continue
c
c       enter input mode
c
```

```
      call reset(table,alias,indx,ni,nr,iu,ru,ip(28),ip(29))
c
c     if any of the integer parameters have changed, we must
c     reinitialize in gdata
c
      do 2 i=1,4
          if(iu(i).ne.1) iu(i)=-1
          if(iu(i).ne.isv(i)) ip(30)=-1
    2 continue
      if(iu(5).eq.0.and.isv(5).ne.0) ip(30)=-1
      if(iu(5).ne.0.and.isv(5).eq.0) ip(30)=-1
      return
      end
```

The display created by *RESET* for this *USRCMD* is:

```
integer parameters

l  left       -1        r  right      -1        i  icont       0
t  top        -1        b  bottom     -1

real parameters

x1 a1x     0.100E+01    bx bux     0.000E+00    c3 cu3     0.000E+00
y1 a1y     0.000E+00    by buy     0.000E+00    cx cexp    0.000E+00
u1 a1u     0.000E+00    c0 cu0     0.100E+01    d0 du0     0.000E+00
x2 a2x     0.000E+00    c1 cu1     0.000E+00    d1 du1     0.000E+00
y2 a2y     0.100E+01    c2 cu2     0.000E+00    e0 eu0     0.000E+00
u2 a2u     0.000E+00
>
```

9.6. Subroutine GDATA

The main input arrays for *PLTMG* or *TRIGEN* are specified through subroutine *GDATA*.

Call GDATA(TITLE, VX, VY, XM, YM, ITNODE, IBNDRY, JB, IP, RP, IU, RU)

A call to *GDATA* is among the first executable statements in *ATEST*. Through this call the user is minimally expected to supply values for *NT*, *NV*, *NC*, and *NB*, (*IP*(1) – *IP*(4)) or *NTR*, *NVR*, *NCR*, and *NBR*, (*IP*(51) – *IP*(54)) as well as the relevant values for the input arrays *TITLE*, *VX*, *VY*, *XM*, *YM*, *ITNODE*, *IBNDRY*, and *JB*. Entries in *RP*, *IU* and *RU*, as well as parameters in *IP* other than those that are required, may be optionally specified in *GDATA*.

Below is an example routine *GDATA*; it defines a mesh on the unit square consisting of eight triangles, nine vertices, eight boundary edges, and no curved edges. Additionally, this *GDATA* interacts with *USRCMD*, allowing one to specify Dirichlet or Neumann boundary conditions independently on each side of the square.

```
        subroutine gdata(title,vx,vy,xm,ym,itnode,
     1      ibndry,jb,ip,rp,iu,ru)
c
            integer itnode(4,*),ibndry(5,*),jb(*),ip(100),iu(100)
            real vx(*),vy(*),xm(*),ym(*),rp(100),ru(100),x(9),y(9)
            character*80 title
            save x,y,irefn,level1,nvtrgt,nt,nv,nc,nb,ispd
c
            data x/0.,0.,.5,1.,1.,1.,.5,0.,.5/
            data y/.5,1.,1.,1.,.5,0.,0.,0.,.5/
            data nt,nv,nc,nb,ispd/8,9,0,8,0/
            data irefn,level1,nvtrgt/2,2,25/
c
c           common /atest2/ibc(4),icont,iu(95),a1x,a1y,a1u,a2x,a2y,
c    1          a2u,bux,buy,cu0,cu1,cu2,cu3,cexp,du0,du1,eu0,ru(84)
c
        title='square'
c
        if(ip(30).eq.1) then
c
c       initialize as laplacian with dirichlet b.c.
c
            do 5 i=1,16
    5           ru(i)=0.0e0
            ru(1)=1.0e0
            ru(5)=1.0e0
            ru(9)=1.0e0
            do 10 i=1,4
   10           iu(i)=-1
            iu(5)=0
        endif
c
c       initialize ip array
c
        ip(1)=nt
        ip(2)=nv
        ip(3)=nc
        ip(4)=nb
        ip(5)=ispd
        ip(6)=nvtrgt
        ip(7)=8
        if(iu(5).ne.0) ip(7)=0
        ip(9)=level1
        ip(10)=irefn
c
c       initialize itnode and ibndry arrays
c       boundary conditions set from using the iu array
c
```

```
      do 20 i=1,nt
          itnode(1,i)=i-1
          itnode(2,i)=i
          itnode(3,i)=9
          itnode(4,i)=0
          if(i.gt.4) itnode(4,i)=1
          ibndry(1,i)=i-1
          ibndry(2,i)=i
          ibndry(3,i)=0
          k=(i+1)/2
          ibndry(4,i)=iu(k)
   20     ibndry(5,i)=i
      itnode(1,1)=8
      ibndry(1,1)=8
c
c     initialize vx and vy arrays
c
      do 30 i=1,nv
          vx(i)=x(i)
   30     vy(i)=y(i)
      return
      end
```

The input data generated by this *GDATA* routine is displayed in figure 9.1.

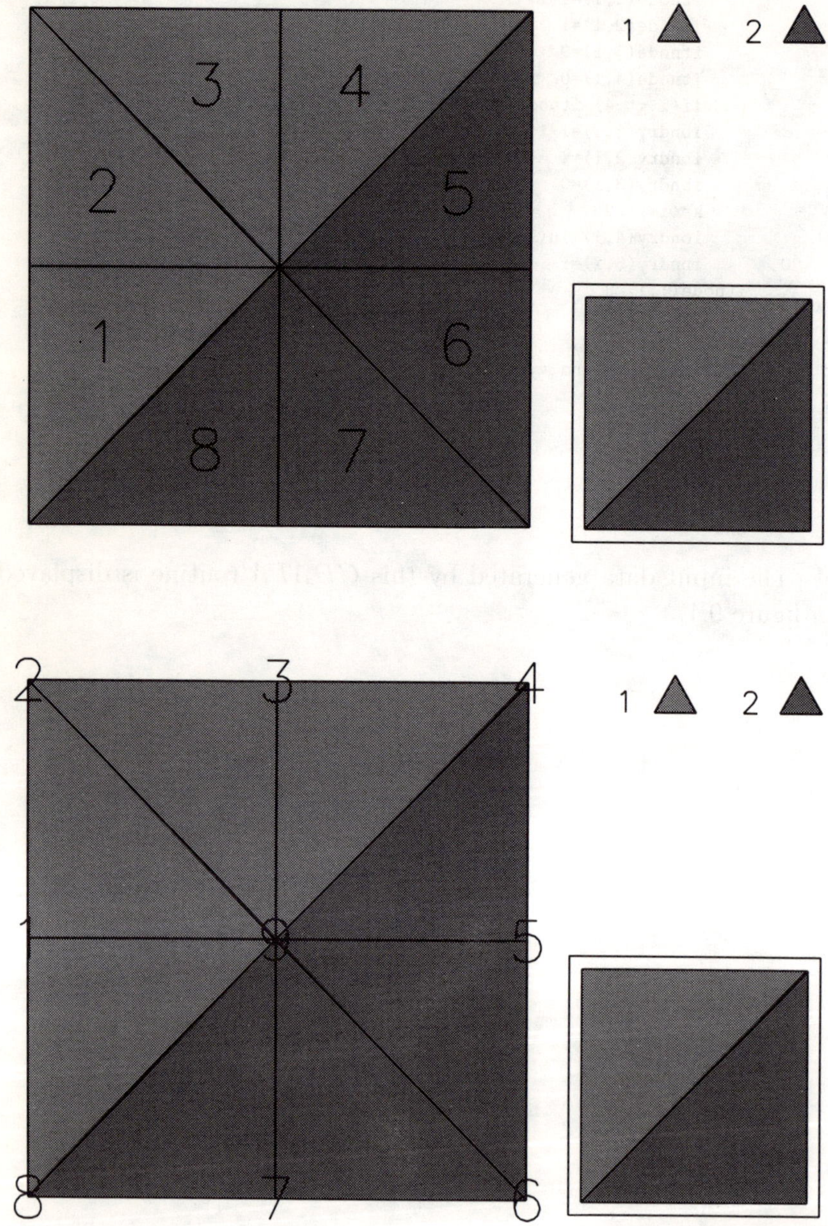

Fig. 9.1. *PLTMG input for the square domain generated by GDATA, with triangle and vertex labels.*

Chapter 10

Machine Dependent Routines

In this chapter, we describe the machine dependent subroutines used by $PLTMG$, $ATEST$, and the graphics routines $TRIPLT$, $GPHPLT$ and $INPLT$.

10.1. Timing

$PLTMG$ and its associated subroutines call the timing function

Function TIMER(I)

Here I is a dummy variable. $TIMER$ is expected to return the time in seconds relative to some fixed time. Time differences are used to determine the time spent in a given routine.

10.2. Graphics

There are three device dependent routines in the graphics package:

Subroutine PLTUTL(NCOLOR, RED, GREEN, BLUE)
Subroutine PLINE(X, Y, N)
Subroutine PFILL(X, Y, N, ICOLOR)

Subroutine $PLTUTL$ takes various actions depending on the value of the integer $NCOLOR$. $NCOLOR > 0$ specifies initialization; $NCOLOR$ will denote the number of colors to be used, and will satisfy

$$2 \leq NCOLOR \leq MXCOLR$$

where $MXCOLR = IP(72)$. RED, $GREEN$, and $BLUE$ are

vectors of length $NCOLOR$. The entries $RED(i)$, $GREEN(i)$, and $BLUE(i)$, $1 \leq i \leq NCOLOR$, are floating point numbers on the interval $[0, 1]$, corresponding to *rgb* values for the *i*th color. Color number 1 is always white ($RED(1) = GREEN(1) = BLUE(1) = 1.0$), and color number 2 is always black ($RED(2) = GREEN(2) = BLUE(2) = 0.0$). The *rgb* values of the remaining entries depend on the picture to be drawn and the value of $MXCOLR$. $PLTUTL$ should make a color map with the required colors, as these will be referenced by future calls from $PFILL$.

If $PLTUTL$ is called with $NCOLOR < 0$, the drawing is complete and any necessary postprocessing should be carried out (e.g., close the plot file). A sample of $PLTUTL$ used for PostScript laser printers is given below.

```
      subroutine pltutl(ncolor,red,green,blue)
c
          real red(*),green(*),blue(*)
          character*8 fname,sname
          common /ps0/iunit,xshift,yshift,scale
          common /ps1/r(256),g(256),b(256)
          save sname,num,fname
          data sname,num/'fig00.ps',1/
c
c     postscript graphics implementation for pltutl
c
      iunit=14
      if(ncolor.lt.0) go to 20
c
c     size is picture size in inches
c
      size=4.0e0
      scale=72.0e0*size
      xshift=72.0e0*(8.5e0-1.5e0*size)/2.0e0
      yshift=72.0e0*(11.0e0-size)/2.0e0
c
      xx=1.5e0*scale+xshift
      yy=scale+yshift
c
c     save color map
c
      do 5 i=1,ncolor
          r(i)=red(i)
          g(i)=green(i)
          b(i)=blue(i)
    5 continue
c
c     create name
c
      ii=num/10
      i0=num-10*ii
```

```
      i1=ii-(ii/10)*10
      fname=sname
      fname(4:4)=char(48+i1)
      fname(5:5)=char(48+i0)
      num=num+1
c
      open(unit=iunit,form='formatted',status='unknown',
     1     file=fname,access='sequential')
      write(iunit,101) xshift,yshift,xx,yy
      return
c
c     print picture
c
   20 write(iunit,102)
      close(unit=iunit)
c
  101 format('%!'/ '%%BoundingBox:',4(2x,f6.1) /
     1  'initgraphics' / '2 setlinejoin'/'1 setlinewidth')
  102 format('showpage')
c
      return
      end
```

Subroutine $PLINE$ has arguments, X, Y, and N. X and Y are vectors of length $N \geq 2$. The points $(X(I), Y(I))$ will lie in the rectangle $(0, 1.5) \times (0, 1)$. The main picture will be in the unit square $(0, 1) \times (0, 1)$, and the two parts of the legend will be drawn in the squares $(1, 1.5) \times (0, .5)$, and $(1, 1.5) \times (.5, 1)$. $PLINE$ is expected to draw the line segments $(X(I), Y(I))$ to $(X(I+1), Y(I+1))$ for $1 \leq I \leq N-1$, rescaling as necessary to device coordinates. A sample of $PLINE$ used for PostScript laser printers is given below.

```
      subroutine pline(x,y,n)
c
          real x(*),y(*)
          common /ps0/iunit,xshift,yshift,scale
          common /ps1/r(256),g(256),b(256)
c
c     postscript graphics implementation for pline
c
      write(iunit,101) r(2),g(2),b(2)
c
      xx=x(1)*scale+xshift
      yy=y(1)*scale+yshift
      write(iunit,102) xx,yy
      do 20 i=2,n
          xx=x(i)*scale+xshift
          yy=y(i)*scale+yshift
          write(iunit,103) xx,yy
   20 continue
c
      write(iunit,104)
```

```
c
  101   format(3(f4.2,2x),'setrgbcolor'/'newpath')
  102   format(f6.1,2x,f6.1,2x,'moveto')
  103   format(f6.1,2x,f6.1,2x,'lineto')
  104   format('stroke')
        return
        end
```

Subroutine $PFILL$ has arguments, X, Y, N, and $ICOLOR$. X and Y are vectors of length $N \geq 3$. The points $(X(I), Y(I))$ will lie in the rectangle $(0, 1.5) \times (0, 1)$, and define a N-sided polygonal region with sides $(X(I), Y(I))$ to $(X(I+1), Y(I+1))$ for $1 \leq I \leq N-1$, and $(X(N), Y(N))$ to $(X(1), Y(1))$. $ICOLOR$ is an integer between 1 and $NCOLOR$, where $NCOLOR$ was the argument that initialized $PLTUTL$, indicating the color to be used. $PFILL$ should color the specified polygon with the specified color. An example $PFILL$ used for PostScript laser printers is given below.

```
        subroutine pfill(x,y,n,icolor)
c
            real x(*),y(*)
            common /ps0/iunit,xshift,yshift,scale
            common /ps1/r(256),g(256),b(256)
c
c       postscript graphics implementation for pfill
c
        write(iunit,101) r(icolor),g(icolor),b(icolor)
c
        xx=x(1)*scale+xshift
        yy=y(1)*scale+yshift
        write(iunit,102) xx,yy
        do 20 i=2,n
            xx=x(i)*scale+xshift
            yy=y(i)*scale+yshift
            write(iunit,103) xx,yy
   20   continue
c
        write(iunit,104)
c
  101   format(3(f4.2,2x),'setrgbcolor'/'newpath')
  102   format(f6.1,2x,f6.1,2x,'moveto')
  103   format(f6.1,2x,f6.1,2x,'lineto')
  104   format('closepath'/ 'fill')
        return
        end
```

If one wishes to use $ATEST$ with more than one device option, one can use the switch $IDEVCE = IP(73)$ to switch among alternatives. The $COMMON$ block $ATEST1$ should be included in $PLTUTL$, $PLINE$, and $PFILL$, and these routines be written allowing output to be directed to one of several devices.

10.3. Obtaining Input

Program $ATEST$ calls subroutine $GETCMD$ to read command strings, and to accept other types of input (e.g., resetting variables) from the user.

Subroutine GETCMD(LIST, PROMPT, ICRTR, ICRTW)

$GETCMD$ is (hopefully) machine independent in most environments. However, it is clearly designed for interactive use, so one may wish to make modifications for a batch environment. $LIST$ is a *character*80* variable, while $PROMPT$ is a *character*1* variable. $ICRTR = IP(28)$ and $ICRTW = IP(29)$ are integers, denoting the standard input and output devices, respectively. In many environments, $ICRTR = 5$ and $ICRTW = 6$. $GETCMD$ itself is a simple routine; it writes the prompt character $PROMPT$, and then reads a command of 80 characters.

```
      subroutine getcmd(list,prompt,icrtr,icrtw)
c
         character*80 list
         character*1 prompt
c
c     get the next input line from
c     the tty or the command file
c
c     print a prompt symbol
c     the $ sign keeps the cursor at the
c     end of the line
c
      write(icrtw,101) prompt
  101 format(1x,a1 $)
c
      read(icrtr,102) list
  102 format(a80)
      return
      end
```

Chapter 11

Examples

In this chapter, we describe in detail the solution of several simple problems using the test driver $ATEST$. The problems are chosen to illustrate many of the features of $PLTMG$, $TRIGEN$, $SKELTN$, $TRIPLT$, $GPHPLT$, and $INPLT$.

11.1. Bratu's Problem

As our first example, we consider the Bratu problem

$$\begin{aligned} -\Delta u - \lambda e^u &= 0 \quad \text{in } \Omega \equiv (0,1) \times (0,1) \\ u &= 0 \quad \text{on } \partial\Omega \end{aligned}$$

and the functional

$$\rho(u, \lambda) = \int_\Omega u^2 dxdy \tag{11.1}$$

The user supplied mesh consists of eight right triangles. The solution manifold for this problem has one limit point. This is illustrated in figure 11.1.

We decide the $LEVEL1$ mesh will be a uniform 5×5 mesh (with 9 non-Dirichlet points). Thus as initialization we specify:

$S\ IR = 2$, $L1 = 2$, $A = 1$, $P = 0$, $R = 0$, $L = 0$

$ATEST$ and $PLTMG$ respond as follows:

```
                level 1 continuation    nv =  25


       lambda        rho     lambda dot   rho dot  determinant eigenvalue
 0  1  0.000E+00  0.000E+00  0.999E+00  0.000E+00  0.411e  9   0.117E+01


iflag =     0          aflag =     0          yflag =     0          cflag =     0
```

The last line of output (the values of all the flags) is printed by $ATEST$; the remainder of the output is printed by $PLTMG$.

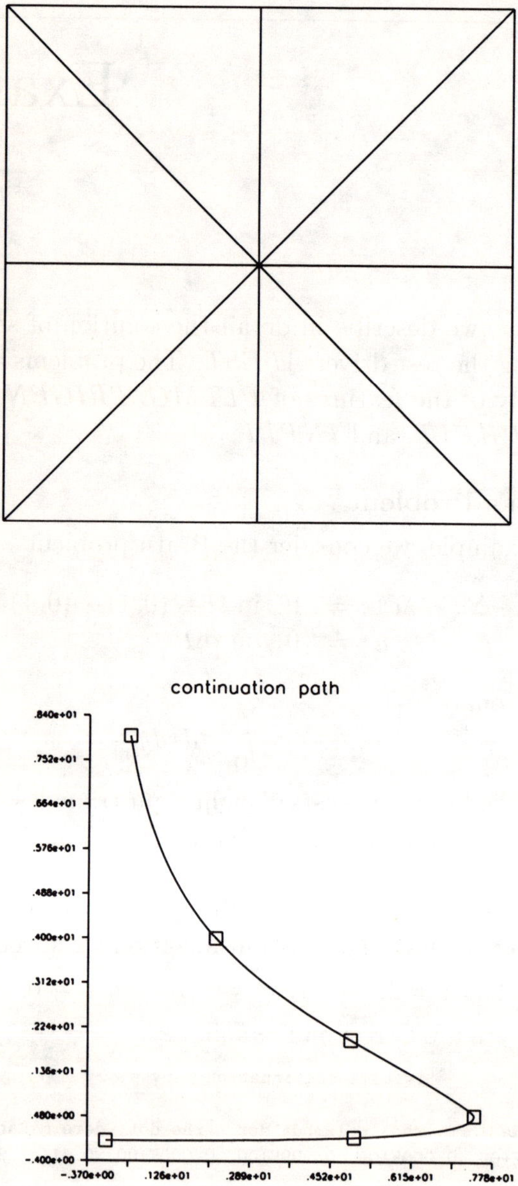

Fig. 11.1. *The initial triangulation and the solution manifold.*

We next decide to continue to $\lambda = 5$.

$S\ RLTRGT = 5$

The result is:

```
                    level 1 continuation    nv =  25

        lambda        rho     lambda dot    rho dot  determinant eigenvalue
 0  3  0.500E+01  0.699E-01  0.997E+00  0.420E-01  0.867e  8   0.779E+00

iflag =    0          aflag =    0          yflag =    0          cflag =    0
```

We now view the solution and its gradient using $TRIPLT$.

$F\ NX = 1, NY = 1, NZ = 1, IFUN = 0$

$F\ NX = 0, NY = 0, NZ = 1, IFUN = 12$

The results are shown in figure 11.2.

We next look for the limit point.

$S\ RTRGT = 2, IPROB = 1$

This results in:

```
                    level 1 continuation    nv =  25

        lambda        rho     lambda dot    rho dot  determinant eigenvalue
 0  4  0.492E+01  0.200E+01  0.982E+00 -0.536E+00 -0.294e  8  -0.187E+01

                    find limit / bifurcation point

 0  3  0.735E+01  0.638E+00 -0.802E+00  0.952E+00 -0.458e  7  -0.174E+00
 0  3  0.741E+01  0.498E+00  0.116E+00  0.140E+01  0.414e  6   0.132E-01
 0  2  0.741E+01  0.508E+00 -0.324E-02  0.142E+01 -0.115e  5  -0.369E-03
 0  1  0.741E+01  0.508E+00 -0.101E-04  0.142E+01 -0.318e  2  -0.115E-05
 0  2  0.741E+01  0.503E+00  0.578E-01  0.141E+01  0.205e  6   0.657E-02
 0  2  0.741E+01  0.508E+00 -0.994E-07  0.142E+01 -0.335e  1  -0.150E-06
 0  2  0.741E+01  0.505E+00  0.289E-01  0.142E+01  0.102e  6   0.328E-02
 0  2  0.741E+01  0.507E+00  0.144E-01  0.142E+01  0.511e  5   0.164E-02
 0  2  0.741E+01  0.508E+00  0.112E-04  0.142E+01  0.402e  2   0.131E-05

iflag =    0          aflag =    0          yflag =    0          cflag =    0
```

In going from $\lambda = 5$ to $\rho = 2$, $PLTMG$ detected a change in the sign of the determinant. Since $IPROB = 1$, $PLTMG$ begins to search for the singular point by the secant/bisection algorithm, solving the equation

$$\dot{\lambda}(\sigma) = 0$$

where σ is the steplength. One can observe superlinear convergence of $\dot{\lambda}$ to zero. The last five steps of the secant/bisection algorithm were required by the convergence criterion for the iteration, which is

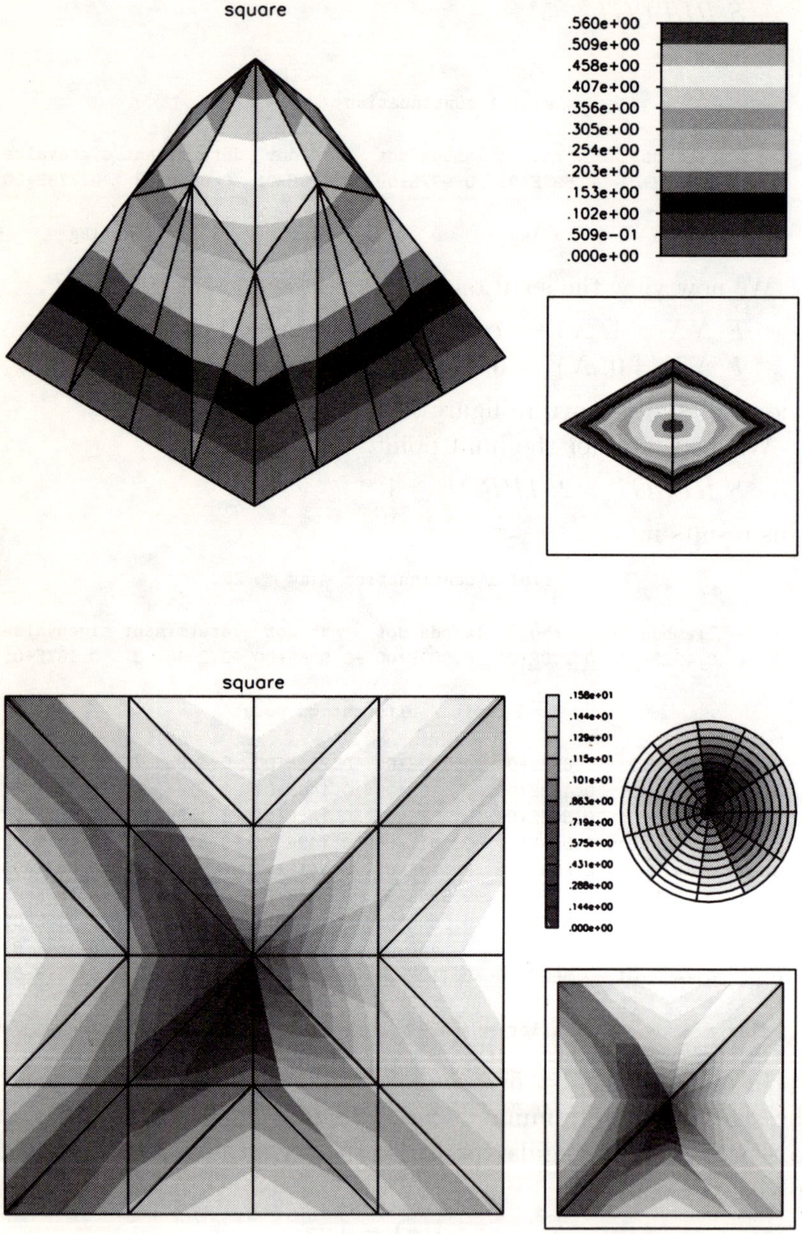

Fig. 11.2. *The solution and its gradient at* $\lambda = 5$.

based on the length of the interval in σ known to contain the root, as well as the size of $\dot{\lambda}$.

At this point, we decide to refine the solution:

$S\ NVTRGT = 100,\ IPROB = 7$

This results in:

```
                 refinement level = 4    nv = 105

          lambda         rho     lambda dot    rho dot   determinant eigenvalue
0  2  0.692E+01  0.508E+00 -0.826E-01  0.142E+01  0.308e  7  -0.248E-02

iflag =     0          aflag =    6          yflag =     0          cflag =     0
```

Here we examine the mesh and the solution.

$F\ IFUN = 0,\ S = 0 * F\ IFUN = -6\ ,\ S = 1$

The resulting pictures are shown in figure 11.3.

We continue the refinement process.

$S\ NVTRGT = 400,\ IADAPT = 3,\ IPROB = 7$

Recall that setting $IADAPT = 3$ allows the refinement to proceed from the current finest mesh ($NV = 105$) rather than restart from the $LEVEL1$ mesh ($NV = 25$).

```
                 refinement level = 5    nv = 401

          lambda         rho     lambda dot    rho dot   determinant eigenvalue
0  2  0.684E+01  0.503E+00 -0.182E-01  0.141E+01  0.411e  7  -0.957E-04

iflag =     0          aflag =    6          yflag =     0          cflag =     0
```

We again examine the mesh and the solution:

$F\ IFUN = 0,\ ISCALE = 0,\ LINES = 2$

$F\ IFUN = 7,\ LINES = 0$

The results are shown in figure 11.4.

We examine the convergence rates, time statistics, and other output provided by $GPHPLT$.

$G13 * G14$

The results are shown in figure 5.2 and 5.3 in chapter 5. Here we see estimated optimal convergence rates for the solution in the $\mathcal{H}^1$ and $\mathcal{L}^2$ norms, indicating the adaptive local mesh refinement procedure is producing good meshes for this problem. We can also observe that the convergence of the hierarchical basis multigrid iteration is not monotonic. This is because at the limit point the matrix is not positive definite; indeed, we essentially have a set of singular, but

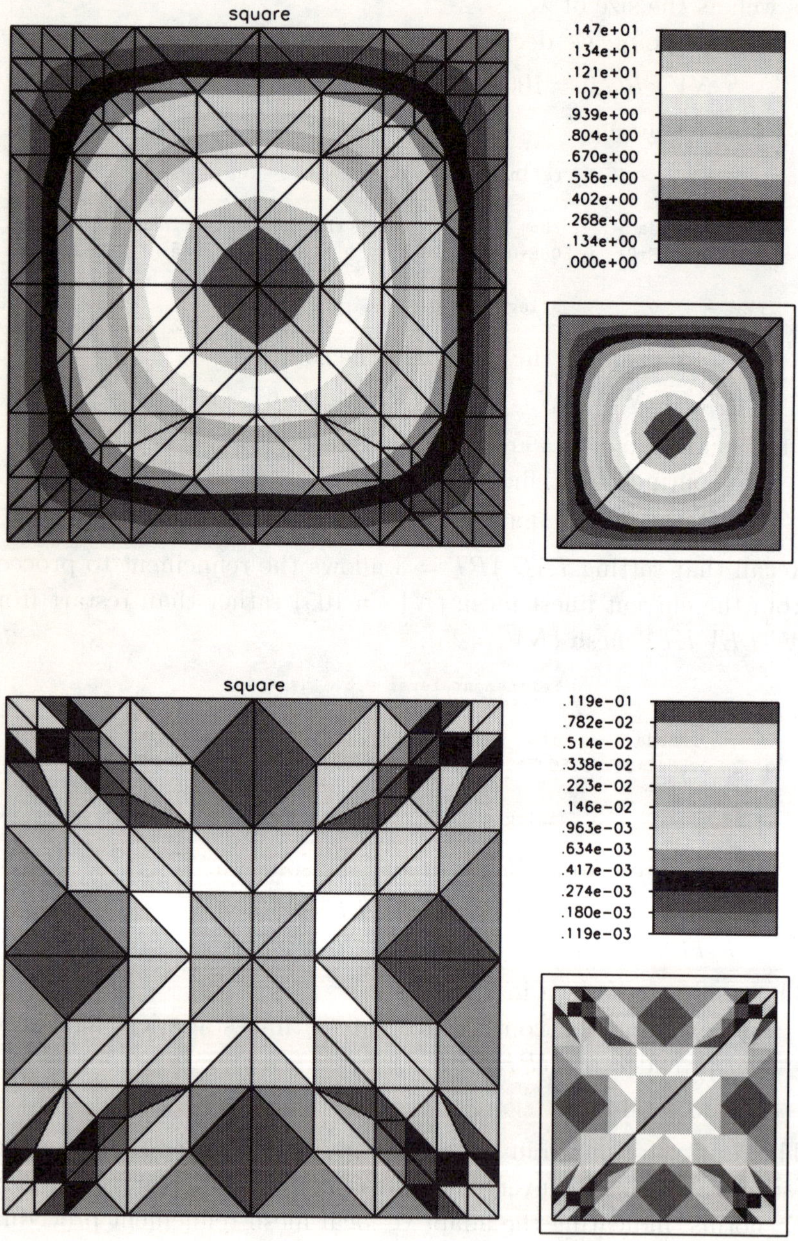

Fig. 11.3. *The solution and the error estimate at the turning point.*

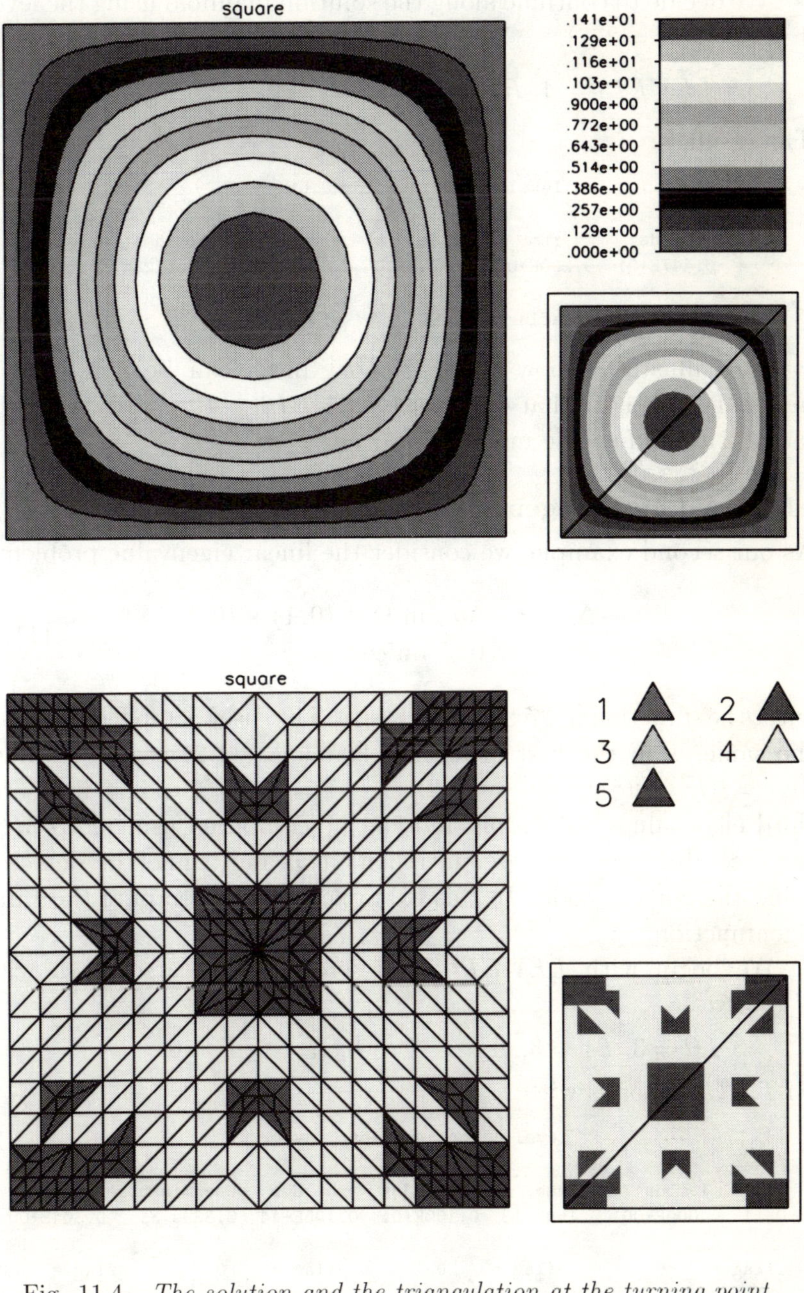

Fig. 11.4. *The solution and the triangulation at the turning point.*

consistent, equations to solve.

We decide to continue along the solution manifold using the level-4 mesh ($NV = 277$).

$$S\ LEVEL1 = 4,\ RTGRT = 2$$

The result is:

```
                 level 1 continuation    nv = 277

          lambda        rho     lambda dot    rho dot   determinant eigenvalue
 0  4  0.447E+01  0.200E+01 -0.979E+00  0.562E+00 -0.598e128   0.796E-01

 iflag =     0          aflag =     0          yflag =     0          cflag =     0
```

This illustrates how the $LEVEL1$ mesh can be redefined, if necessary. We note that our choice $LEVEL1 = 4$ requires the LDU factorization of sparse matrices of order 277.

11.2. A Linear Eigenvalue Problem

As our second example, we consider the linear eigenvalue problem

$$\begin{aligned} -\Delta u &= \lambda u \quad \text{in } \Omega \equiv (0,1) \times (0,1) \\ u &= 0 \quad \text{on } \partial\Omega \end{aligned} \tag{11.2}$$

with the functional given in (11.1). The user supplied mesh is the same as in the first example. Problem (11.2) has eigenvalues $\lambda_{k\ell} = (k^2 + \ell^2)\pi^2$, $k = 1, 2, \ldots$, $\ell = 1, 2, \ldots$. The first and third eigenvalues are simple and can be computed using $PLTMG$; the second eigenvalue has multiplicity two and cannot be detected using the current code. In this example, we will compute the third eigenfunction.

We begin with $LEVEL1 = 3$, creating a coarse mesh with $NV = 81$.

$$S\ IR = 3,\ L1 = 3,\ A = 1,\ P = 0,\ R = 0,\ L = 0$$

$PLTMG$ responds with:

```
                 level 1 continuation    nv =  81

          lambda        rho     lambda dot    rho dot   determinant eigenvalue
 0  1  0.000E+00  0.159E-13  0.100E+01  0.105E-14  0.339e 35   0.304E+00

 iflag =     0          aflag =     0          yflag =     0          cflag =     0
```

Next we continue to $\lambda = 30$, beyond the first eigenvalue.

$$S\ RLTRGT = 30$$

This results in:

```
                        level 1 continuation    nv =  81

          lambda        rho    lambda dot   rho dot  determinant eigenvalue
  0  2  0.300E+02  0.267E-27  0.100E+01 -0.147E-28 -0.106e 32  -0.146E+00

  iflag =     0         aflag =    0         yflag =    0         cflag =    0
```

Notice that both the determinant and the smallest singular value μ change sign; however, $\dot{\lambda}$ does not change sign. We continue to $\lambda = 70$, past the eigenvalue of multiplicity 2.

$S\ RLTRGT = 70$

This results in:

```
                        level 1 continuation    nv =  81

          lambda        rho    lambda dot   rho dot  determinant eigenvalue
  0  1  0.700E+02  0.219E-26  0.100E+01 -0.107E-27 -0.229e 28   0.550E+00

  iflag =     0         aflag =    0         yflag =    0         cflag =    0
```

Note that the determinant does not change sign; $PLTMG$ could not detect the double eigenvalue. We now look for the third eigenvalue:

$S\ IPROB = 1,\ RLTRGT = 100$

This results in:

```
                        level 1 continuation    nv =  81

          lambda        rho    lambda dot   rho dot  determinant eigenvalue
  0  1  0.100E+03  0.157E-26  0.100E+01 -0.580E-28  0.231e 25  -0.109E+00

                        find limit / bifurcation point

  0  1  0.925E+02  0.433E-28  0.100E+01 -0.270E-28  0.238e 25  -0.135E-01
  0  1  0.900E+02  0.763E-28  0.100E+01  0.106E-27 -0.585e 25   0.184E-01
  0  1  0.915E+02  0.761E-28  0.100E+01  0.435E-23 -0.158e 21   0.437E-06
  0  1  0.915E+02  0.311E-25  0.100E+01 -0.121E-21  0.142e 22  -0.663E-05

  iflag =     0         aflag =    0         yflag =    0         cflag =    0
```

To find the bifurcation point, the secant/bisection algorithm solves the equation

$$\mu(\sigma) = 0$$

where σ is the steplength. Again we observe superlinear convergence.

We next switch branches at the bifurcation point:

$S\ IPROB = 2$

The response is:

```
                  level 1 continuation    nv =  81

        lambda        rho     lambda dot    rho dot  determinant eigenvalue
 0  0  0.915E+02  0.311E-25  0.000E+00  0.113E-02  0.142e 22  -0.663E-05

 iflag =     0          aflag =     0          yflag =     0          cflag =     0
```

In this call to $PLTMG$, a new tangent vector for the bifurcating branch has been computed; the solution itself remained unchanged (in this case, it is the trivial solution). We can tell that the branch switch has occurred by noting the change in value of $\dot{\lambda}$ from 1 to 0. The next continuation step will be along the bifurcating branch. We decide to compute an eigenvector with unit $\mathcal{L}^2$ norm:

$S\ RTRGT = 1$

The response is:

```
                  level 1 continuation    nv =  81

        lambda        rho     lambda dot    rho dot  determinant eigenvalue
 0  5  0.915E+02  0.121E-01  0.154E-03  0.220E+00 -0.677e 20   0.196E-06
 0  4  0.915E+02  0.100E+01  0.247E-04  0.200E+01 -0.559e 20   0.239E-06

 iflag =     0          aflag =     0          yflag =     0          cflag =     0
```

Here $PLTMG$ took two steps to reach the target value in ρ. After switching branches, both $\dot{\lambda}$ and $\dot{\rho}$ were considered to be zero by $PLTMG$. In response, it switched normalization functions from (2.4) to (2.5) for one step [7]. Once we compute a point a small distance away from the bifurcation point on the bifurcating branch, the standard normalization function becomes well defined once more (note $\dot{\rho}$ is no longer close to zero), and we can then compute the target point in one step.

At this point we choose to refine the mesh, in order to compute a more accurate eigenvalue and eigenfunction. We want the $\mathcal{L}^2$ norm to remain constant.

$S\ IPROB = 6,\ NVTRGT = 300$

The response is:

```
                  refinement level = 5    nv = 305

        lambda        rho     lambda dot    rho dot  determinant eigenvalue
 0  2  0.819E+02  0.100E+01 -0.135E-02  0.200E+01 -0.197e 27  -0.509E-05

 iflag =     0          aflag =     6          yflag =     0          cflag =     0
```

We next look at convergence rates and timing statistics

$G\ IGRSW = 13$

yielding figure 11.5.

Here we observe good estimated rates of convergence for the solution, indicating the adaptive refinement procedure has worked well. The nonmonotonic convergence of the hierarchical basis multigrid iteration is caused by an indefinite stiffness matrix. Next, we make pictures of the solution and the mesh.

$F\ IFUN = 7$

$F\ IFUN = 0,\ NX = -1,\ NY = 1,\ NZ = 1,\ LS = 2$

The results are shown in figure 11.6.

11.3. A Convection-Diffusion Equation

Our third example is the convection-diffusion equation

$$\begin{aligned} -\Delta u - \beta^t \nabla u &= 1 \quad \text{in } \Omega \equiv (0,1) \times (0,1) \\ u &= 0 \quad \text{on } \partial\Omega \end{aligned}$$

where $\beta = (100, 0)^t$. The user supplied mesh is the same as in the first example. This problem will not be solved as a continuation problem, so $IPROB = 8$.

We will begin by adaptively solving the problem, creating a mesh with about 400 vertices.

$S\ IREFN = 2,\ LEVEL1 = 2,\ NVTRGT = 400$

$PLTMG$ responds with:

```
                  refinement level = 2    nv =  25

         lambda       rho     lambda dot    rho dot   determinant eigenvalue
 0  1  0.000E+00  0.145E-04  0.000E+00  0.000E+00  0.158e 28   0.000E+00

                  refinement level = 5    nv = 109

         lambda       rho     lambda dot    rho dot   determinant eigenvalue
 0  2  0.000E+00  0.181E-04  0.000E+00  0.000E+00  0.203e 25   0.000E+00

                  refinement level = 7    nv = 407

         lambda       rho     lambda dot    rho dot   determinant eigenvalue
 0  2  0.000E+00  0.234E-04  0.000E+00  0.000E+00  0.580e 21   0.000E+00

iflag =      0         aflag =     6         yflag =     0         cflag =     0
```

We can examine the mesh and the solution.

$F7$

$F\ IFUN = 0,\ NX = -1,\ NY = -1,\ NZ = 1,\ LS = 2$

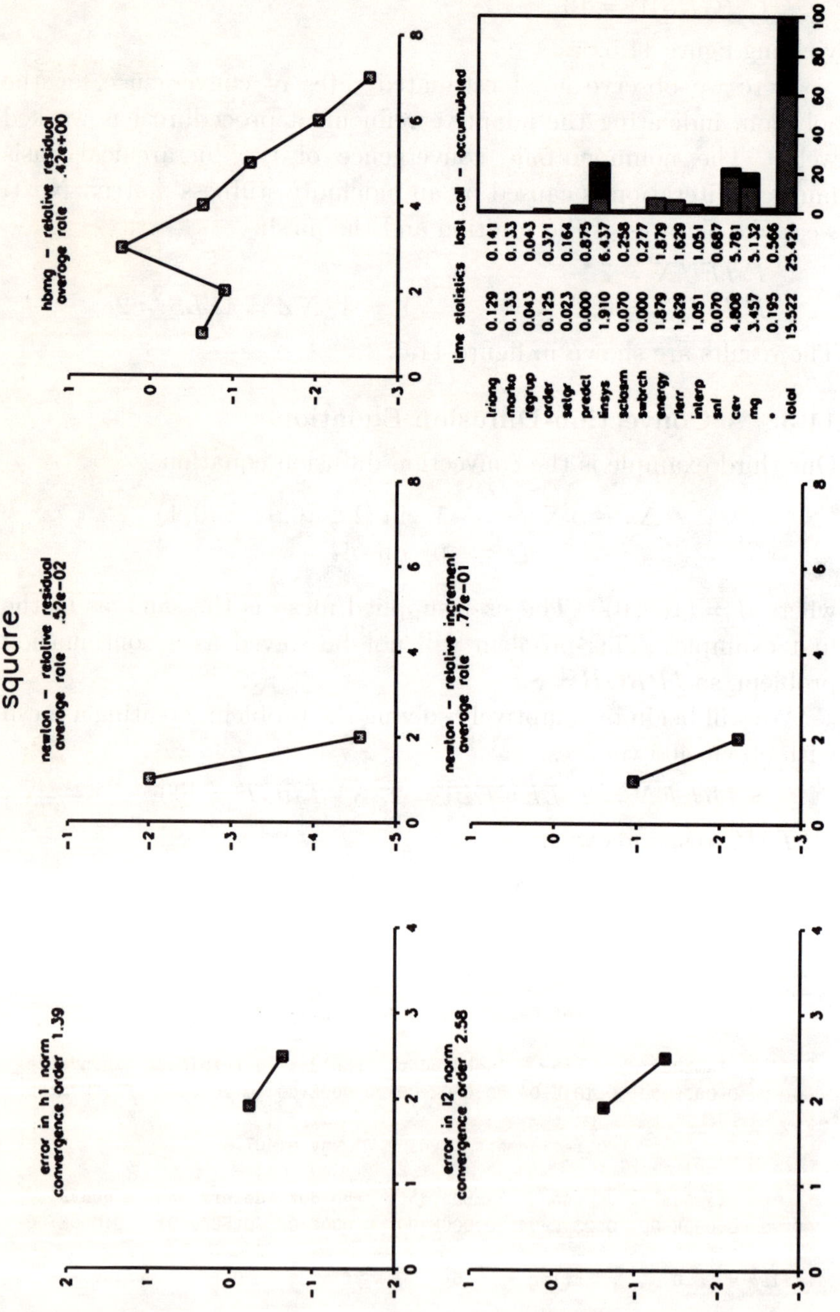

Fig. 11.5. *GPHPLT output for IGRSW* = 13.

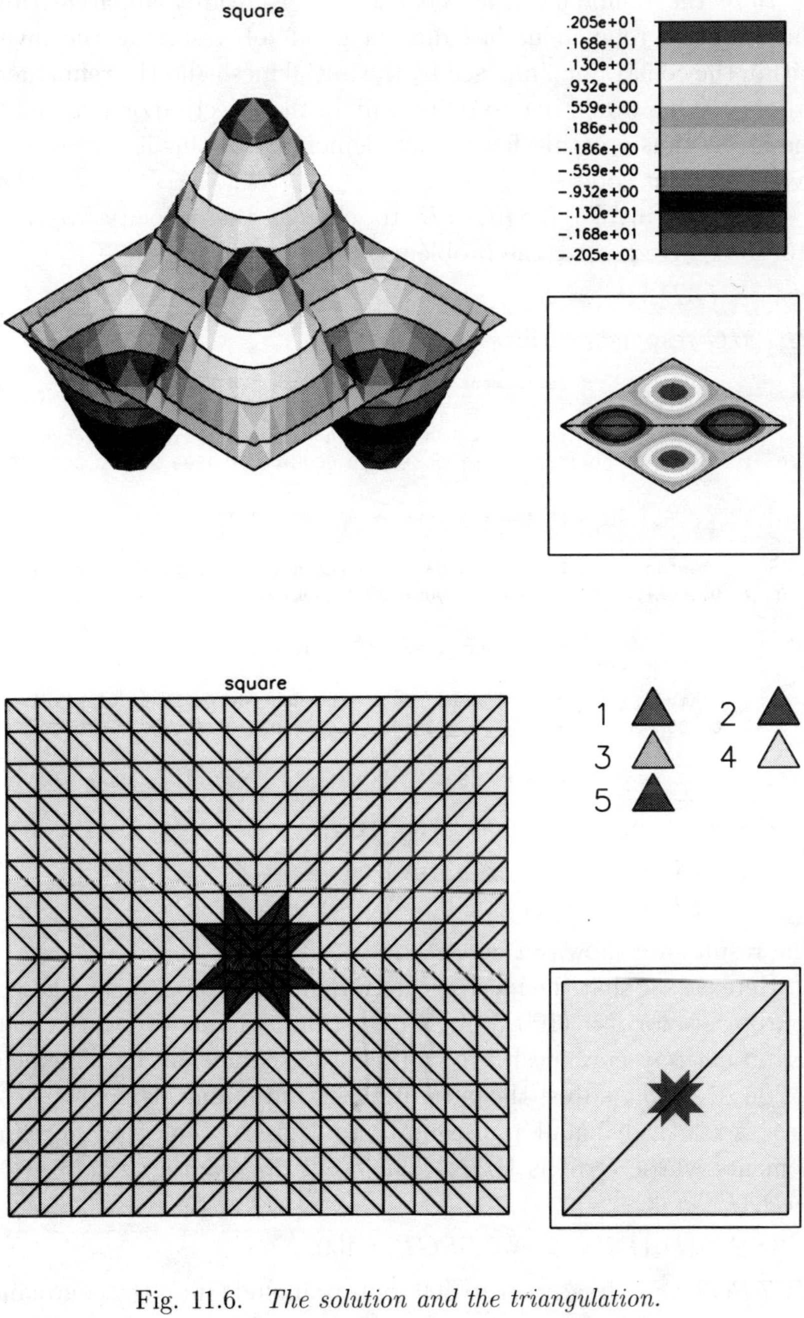

Fig. 11.6. *The solution and the triangulation.*

The results are shown in figure 11.7.

Here the boundary layer near $x = 0$ is readily apparent, and the adaptive refinement has done a good job resolving the layer, within the constraints imposed by the initial mesh and the refinement process. (One would like to refine only in the direction orthogonal to the layer, thus creating fewer thin elements, but this is not allowed by the current code).

We now invoke $USRCMD$ to change the velocity to $\beta = (100, 50)^t$. We resolve the problem on the current mesh.

$S\ IADAPT = 2$

$PLTMG$ responds with:

```
                    refinement level = 2    nv =   25

          lambda         rho     lambda dot    rho dot  determinant eigenvalue
   0  1  0.000E+00  0.128E-04  0.000E+00  0.000E+00  0.158e 21   0.000E+00

                    refinement level = 4    nv =  111

          lambda         rho     lambda dot    rho dot  determinant eigenvalue
   0  2  0.000E+00  0.187E-04  0.000E+00  0.000E+00  0.149e 20   0.000E+00

                    refinement level = 7    nv =  407

          lambda         rho     lambda dot    rho dot  determinant eigenvalue
   0  2  0.000E+00  0.198E-04  0.000E+00  0.000E+00  0.116e 20   0.000E+00

 iflag =    0          aflag =    8          yflag =     0          cflag =    0
```

We graph the solution and the error.

$F\ IFUN = 0,\ NX = -1,\ NY = -1,\ NZ = 1,\ LS = 2$
$F\ IFUN = -6,\ NX = 0,\ NY = 0,\ LS = 0,\ S = 1$

The results are shown in figure 11.8.

Here we see that the mesh created when $\beta = (100, 0)^t$ is no longer appropriate for $\beta = (100, 50)^t$. Rather than starting over and creating an entirely new mesh, we can use the refinement/unrefinement option to simply adjust the current mesh, unrefining elements whose error is relatively small and using these degrees of freedom to refine elements whose error is large. This is all done automatically with $IADAPT = 3$.

$S\ IADAPT = 3,\ NVTRGT = 400$

$NVTRGT$ was already set to 400, but we include it on the command line to emphasize that we start from a mesh with about 400 vertices, and end with a different mesh with about the same number of

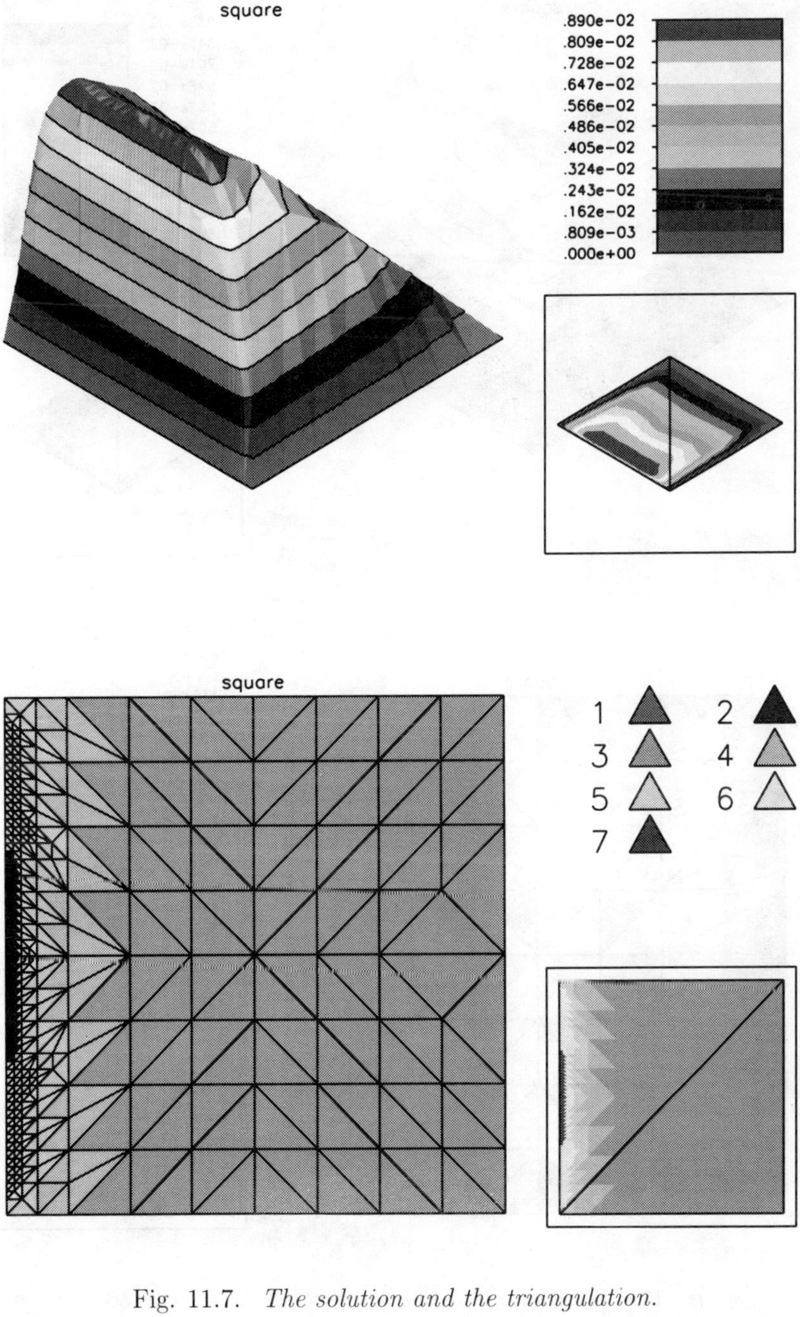

Fig. 11.7. *The solution and the triangulation.*

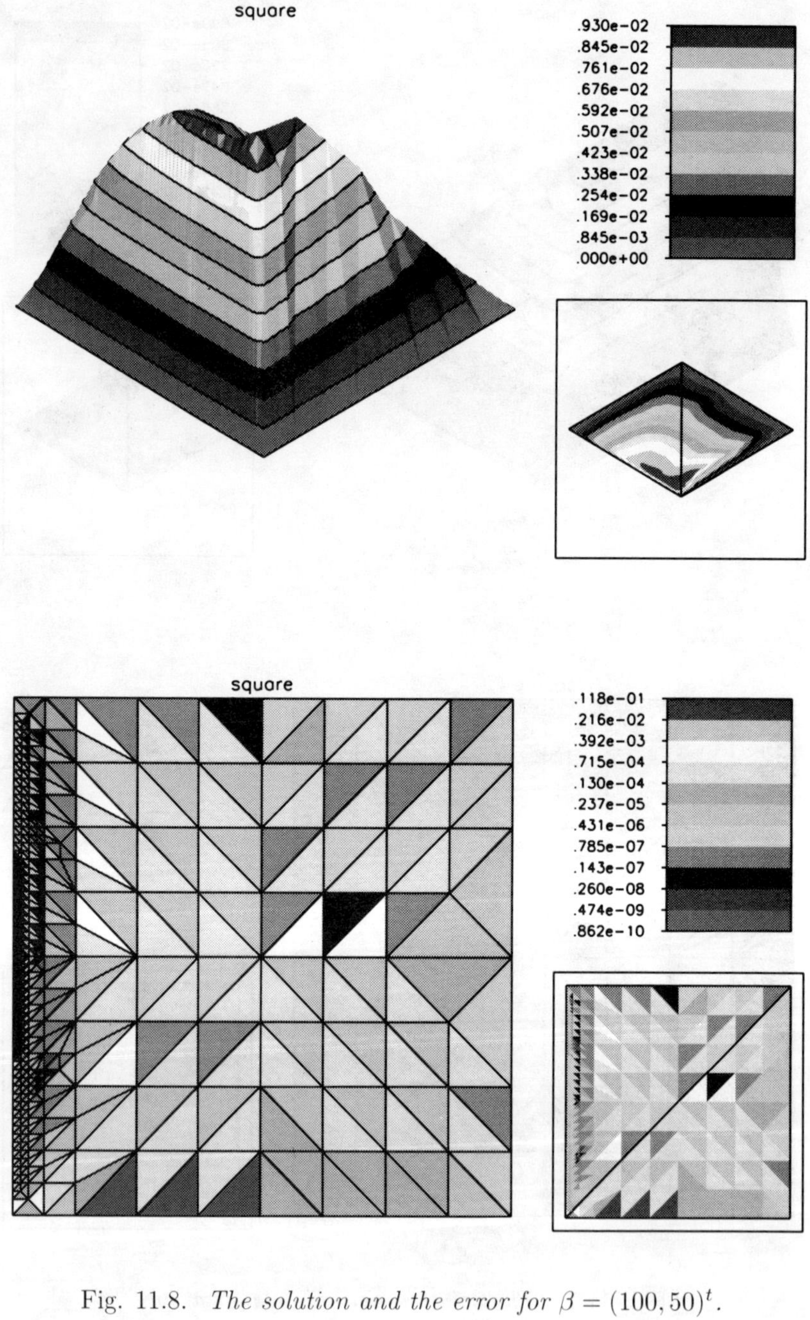

Fig. 11.8. *The solution and the error for* $\beta = (100, 50)^t$.

vertices.

```
                  refinement level = 7    nv = 407

        lambda       rho     lambda dot    rho dot   determinant eigenvalue
0  2  0.000E+00  0.199E-04  0.000E+00  0.000E+00  0.540e 19   0.000E+00

iflag =    0        aflag =    6         yflag =    0          cflag =    0
```

We again examine the mesh and the solution.

$F\ F = 0,\ NX = -1,\ NY = -1,\ NZ = 1,\ LS = 2,\ S = 0$

$F\ IFUN = 7,\ NX = 0,\ NY = 0,\ LINES = 0$

The results are shown in figure 11.9.

We can see that the new mesh has refined elements near the origin along the line $y = 0$, and unrefined along the line $x = 0$ near $y = 1$. Overall, this mesh is more appropriate for the changed character of the boundary layer.

11.4. Creating a Mesh Using TRIGEN

In this example, we want to create an input triangulation for $PLTMG$ using $TRIGEN$. The region we wish to triangulate is pictured in figure 11.10.

This region consists of $NTR = 6$ subregions, using $NVR = 30$ vertices, $NBR = 35$ edges, none curved ($NCR = 0$). We will experiment with different values of the parameters $HMAX$ and $GRADE$ to illustrate their influence on the resulting mesh. We begin with the command:

$T\ HMAX = 0,\ GRADE = 1.5$

$TRIGEN$ responds with:

```
oflag =    0     nt   =  475     nv   =  276     nc   =    0     nb   =   75
```

This output line is printed in $ATEST$, indicating $TRIGEN$ has successfully computed a mesh with 475 triangles and 276 vertices. We can get some indication of the quality of the mesh using:

$F9$

calling $TRIPLT$ to illustrate the mesh, with the elements colored according to the $TRIGEN$ quality measure (3.1). The result is given in figure 11.11.

An I command could also be used here, or $F8$ to view the mesh by region, $F10$ to view the mesh by maximum element angles, or $F11$ to view the mesh by minimum element angles.

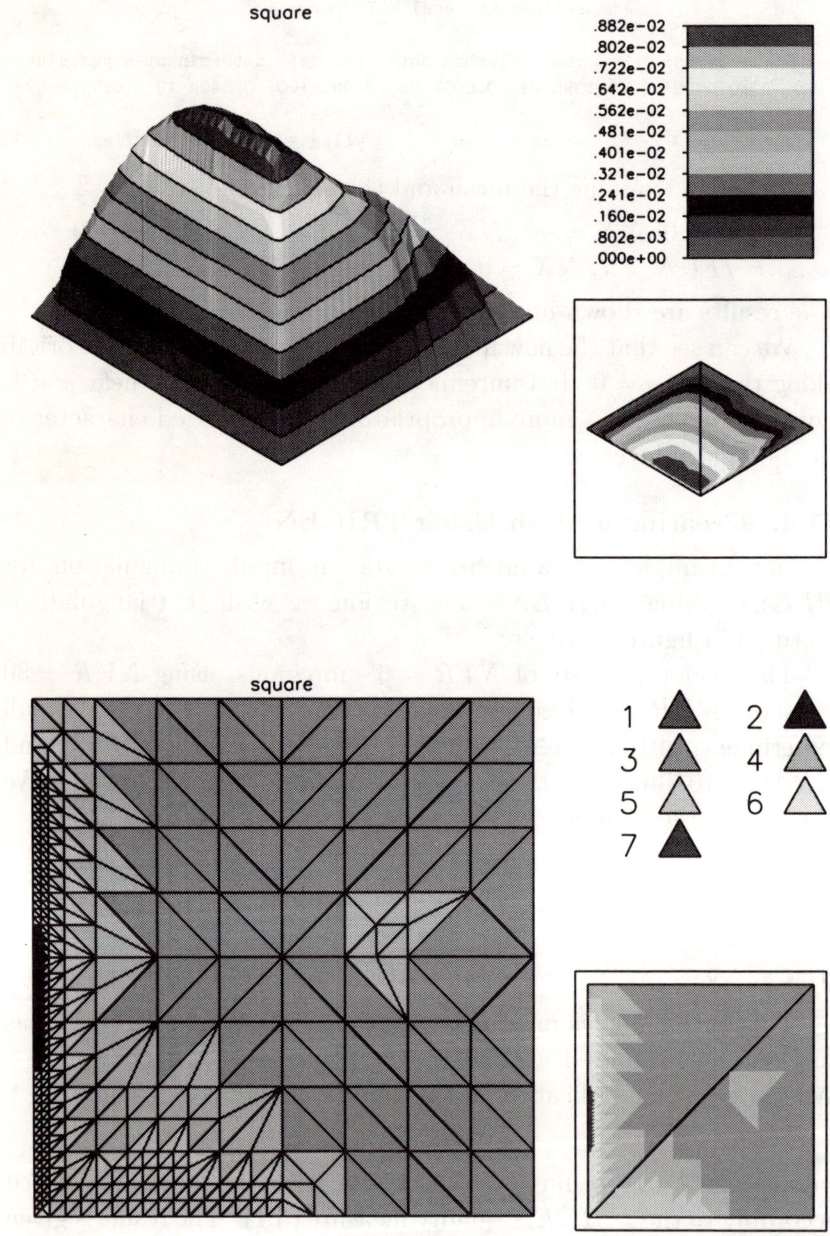

Fig. 11.9. *The solution and the adjusted triangulation.*

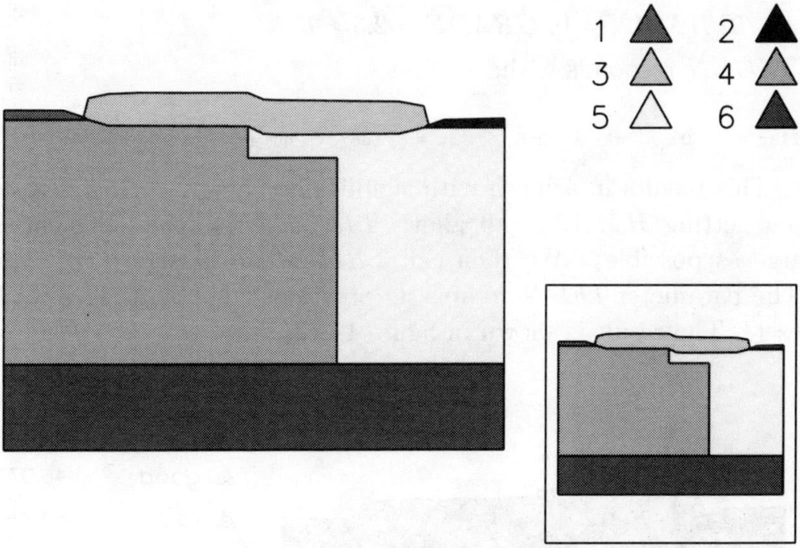

Fig. 11.10. *A sample region for TRIGEN.*

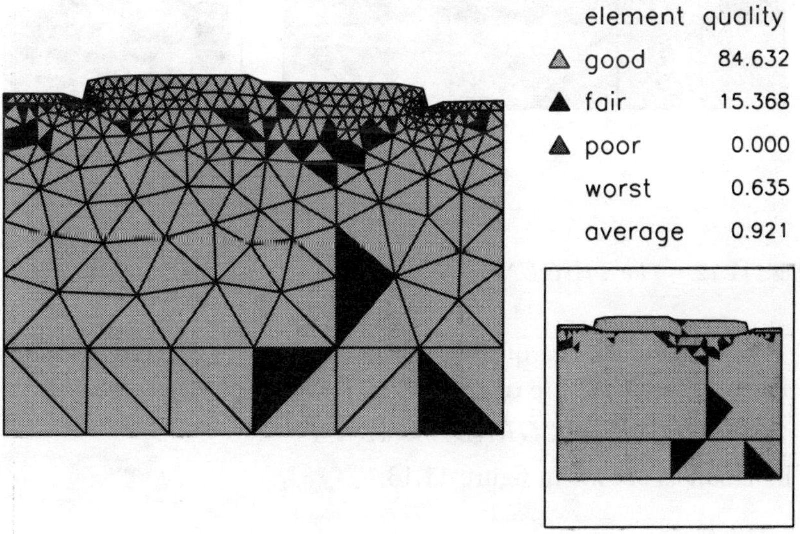

Fig. 11.11. *The TRIGEN mesh with $HMAX = 0.$ and $GRADE = 1.5$.*

We next illustrate the effect of the parameter $GRADE$.

$T\ HMAX = 0,\ GRADE = 2.5 * F$

$TRIGEN$ responds with:

```
oflag =    0     nt  =  265     nv =   165     nc =    0     nb  =   63
```

This results in a mesh with significantly fewer vertices. (Recall that setting $HMAX = 0$ allows $TRIGEN$ to make elements as large as possible). We then call $TRIPLT$ to illustrate the mesh. (The parameter $IFUN$ retains its previous value, and need not be reset). The result is shown in figure 11.12.

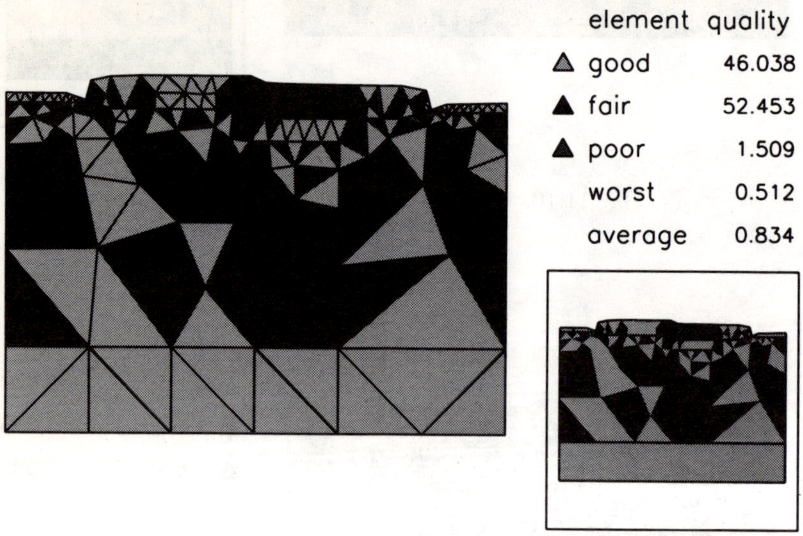

Fig. 11.12. *The $TRIGEN$ mesh with $HMAX = 0.$ and $GRADE = 2.5$.*

As we can see, the quality of this mesh is lower. We can obtain a mesh of intermediate quality with, for example:

$T\ HMAX = 0,\ GRADE = 1.7 * F$

The result is shown in figure 11.13.

```
oflag =    0     nt  =  392     nv  =  233     nc  =    0     nb  =   72
```

To illustrate the effect of $HMAX$ we try, for example:

$T\ HMAX = .02,\ GRADE = 1.5 * F$

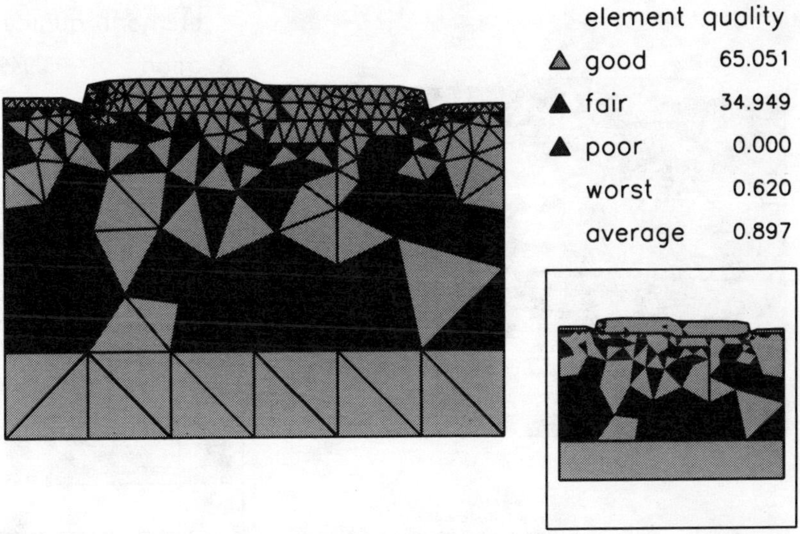

Fig. 11.13. *The TRIGEN mesh with* $HMAX = 0.$ *and* $GRADE = 1.7.$

This yields figure 11.14.

```
oflag =    0    nt =   975    nv =   542    nc  =    0    nb  =  107
```

The size of $HMAX$ depends of the scaling of the (x, y) coordinates provided in the arrays VX and VY. As we can see, the triangles are of more uniform size, and the triangulation is of high quality as measured by (3.1). On the other hand, this is already a reasonably fine coarse mesh, and $PLTMG$ would have to solve the linear systems generated by this mesh by sparse Gaussian elimination. Generally, one must make a compromise in selecting a coarse mesh from among these possibilities (and other possible meshes resulting from other settings of $HMAX$ and $GRADE$).

11.5. Creating a Skeleton Using SKELTN

In this example, we create an input skeleton for $TRIGEN$ using $SKELTN$. To do this, we must first solve the problem on a crude mesh using $PLTMG$. In this example, we will solve the equation

$$-\Delta u = 1$$

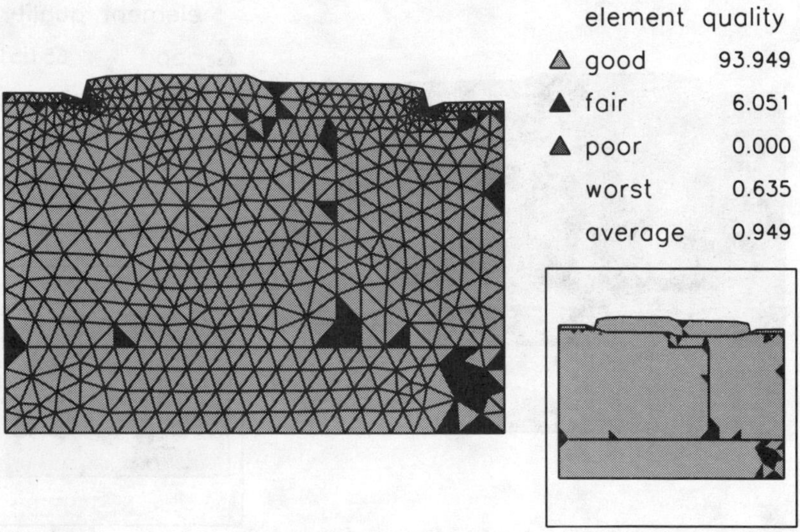

Fig. 11.14. *The TRIGEN mesh with* $HMAX = .02$ *and* $GRADE = 1.5$.

with homogeneous Dirichlet boundary conditions on the unit square $(0, 1) \times (0, 1)$.

We will begin with a simple mesh consisting of eight triangles and nine vertices illustrated in figure 11.15.

To generate a solution on a coarse mesh we first call $PLTMG$ in

$$S\ IREFN = 2,\ LEVEL1 = 2 * F0 * D5$$

With these settings for $IREFN$ and $LEVEL1$, $PLTMG$ will generate a solution on a uniform 5×5 mesh. The result is:

```
                      refinement level = 2   nv =  25

        lambda        rho     lambda dot   rho dot  determinant eigenvalue
 0  1  0.000E+00  0.145E-02  0.000E+00  0.000E+00  0.411e  9   0.000E+00

  iflag =     0         aflag =     6         yflag =     0         cflag =     0
```

We then draw the resulting solution using $TRIPLT$. The result is illustrated in figure 11.15. The command $D5$ saves the solution in the file $RW105.SV$. This is required since calls to $TRIGEN$ will overwrite the solution in the W array; since we plan to create several skeletons, we must periodically restore the solution.

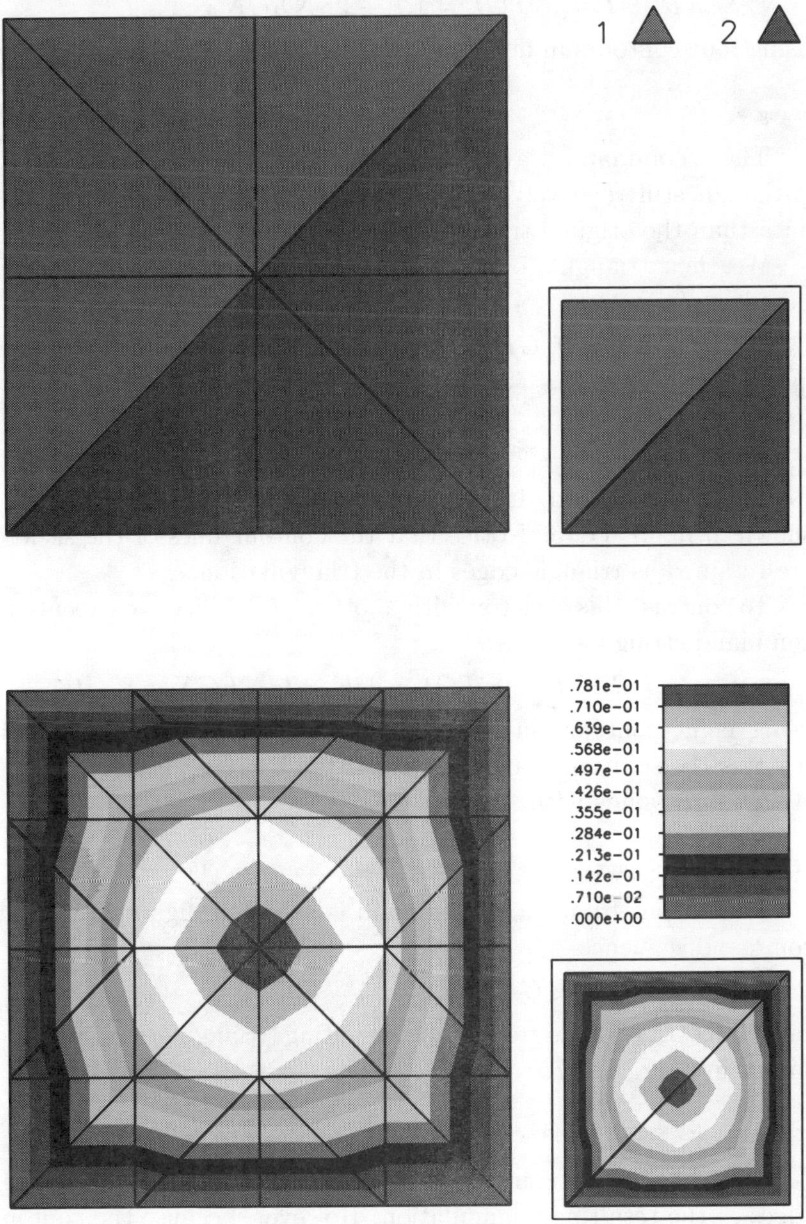

Fig. 11.15. *The initial triangulation, top, and the initial solution.*

We first call $SKELTN$ with

$$K\ FRACT = 0,\ TOL = 1.E-2,\ NRGN = 4 * I$$

using four contours in defining the skeleton. $ATEST$ responds with

```
oflag =    0     ntr =    8     nvr =   24     ncr =    8     nbr =   31
```

The I command draws the input, that is shown in figure 11.16. Although still relatively simple, this skeleton requires more input data than the original triangulation.

We then triangulate the region using $TRIGEN$ via the commands

$$T\ HMAX = 0,\ GRADE = 1.5 * F9$$

The result is

```
oflag =    0     nt  =  280     nv  =  161     nc  =    0     nb  =   40
```

The F command draws the resulting triangulation, which is shown in figure 11.16. Notice that the contour lines of the skeleton are retained as triangle edges in the triangulation.

To contrast this skeleton with another possibility, we execute the command string

$$L * K\ FRACT = 0,\ TOL = 1.E-2,\ NRGN = 8 * I$$

using eight contours in defining the skeleton. The L command restores the solution that was overwritten by the call to $TRIGEN$. $ATEST$ responds with

```
oflag =    0     ntr =   16     nvr =   64     ncr =   16     nbr =   79
```

The output from the I command is shown in figure 11.17. The command sequence

$$T\ HMAX = 0,\ GRADE = 1.5 * F9$$

calls $TRIGEN$ and graphs the resulting triangulation, which is shown in figure 11.17.

```
oflag =    0     nt  = 1384     nv  =  733     nc  =    0     nb  =   80
```

Once again, the contour lines in the skeleton become triangle edges in the resulting triangulation. However, because the contours lines are more closely spaced in the case $NRGN = 8$, more triangles are created than for the case $NRGN = 4$. As with $TRIGEN$, one may have to experiment with different parameter settings in $SKELTN$ to find the best skeleton.

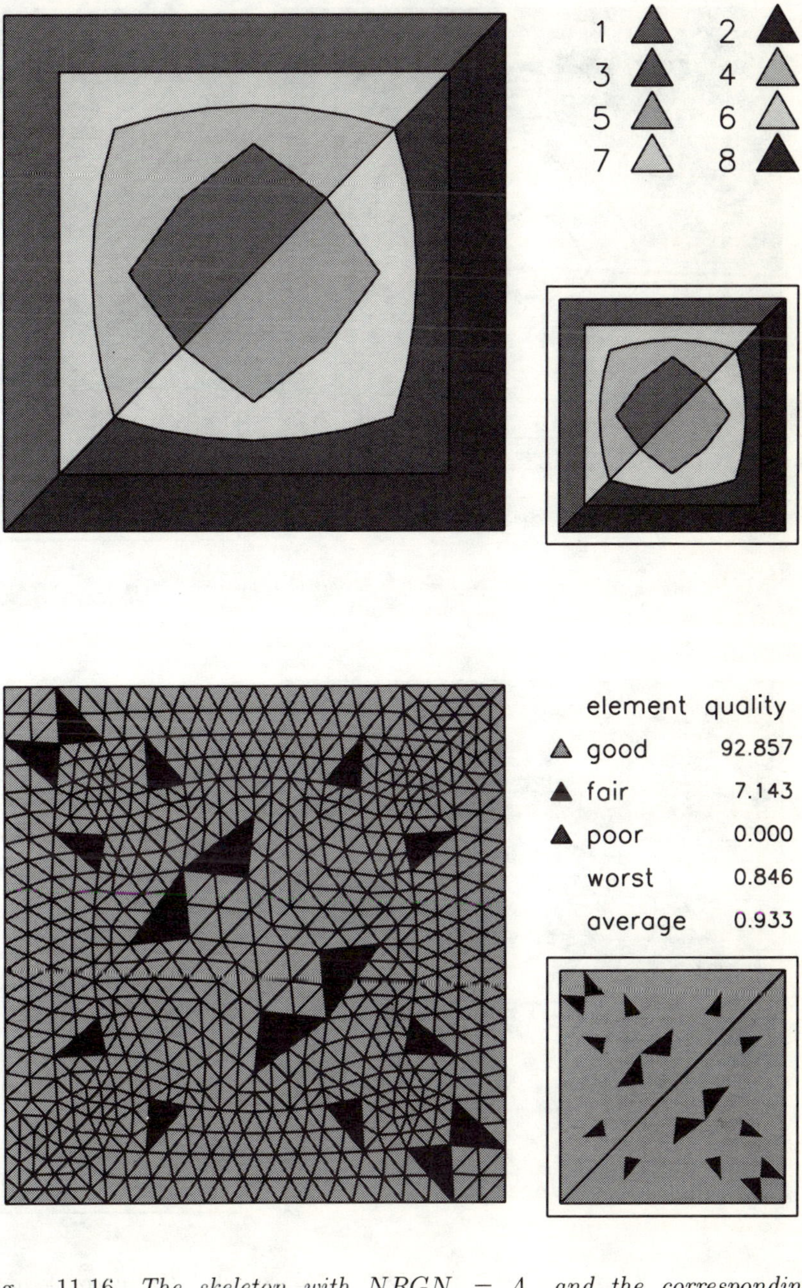

Fig. 11.16. *The skeleton with NRGN = 4, and the corresponding triangulation.*

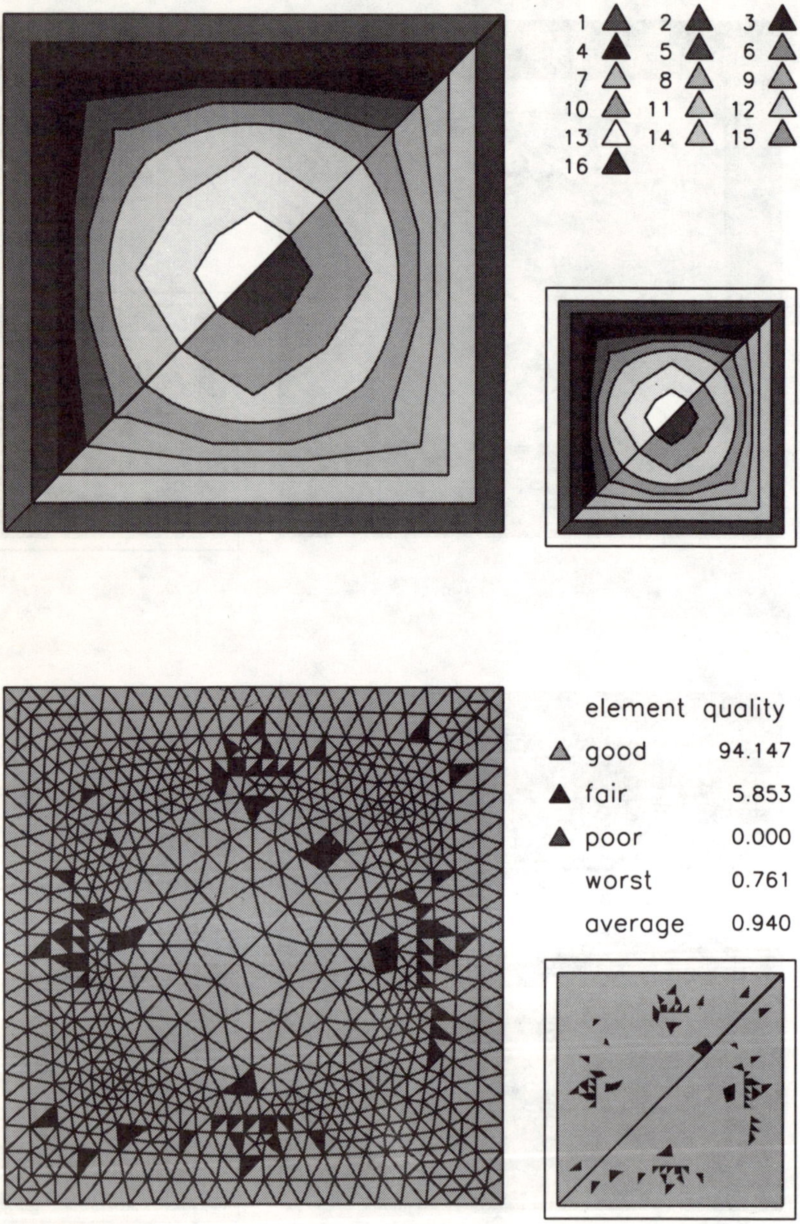

Fig. 11.17. *The skeleton with NRGN = 8, and the corresponding triangulation.*

Chapter 12

Test Problems

In this chapter we briefly document the test problem data sets included with the $PLTMG$ source code. These problems encompass a variety of applications, and exercise most features of the package. Each data set minimally consists of functions $A1XY$, $A2XY$, FXY, GXY, $P1XY$, $P2XY$, QXY, and subroutines $USRCMD$ and $GDATA$. Problem specific routines are also included.

12.1. Test Problem CIRCLE

In this example, we solve the Laplace equation

$$-\Delta u = 0$$

in the unit circle, with a crack along the positive x axis. Homogeneous Dirichlet boundary conditions are imposed on the top of the crack, and homogeneous Neumann boundary conditions are imposed below the crack. On the boundary of the circle, nonhomogeneous boundary conditions are imposed such that the true solution is given by

$$u = r^{\alpha} \sin \alpha\theta \tag{12.1}$$

where $\alpha = 1/4$. This is the leading term of the singularity arising from the interior angle of 2π coupled with the change of boundary conditions at the origin.

The domain is defined by a triangulation consisting of eight similar triangles, so $TRIGEN$ is not needed to create an initial mesh.

The $USRCMD$ for this test problem has two parameters that can be altered. IBC can be set to ± 1. If $IBC = -1$ the boundary conditions on the outer boundary of the circle are nonhomogeneous

Dirichlet as described above; if $IBC = 1$, nonhomogeneous Neumann boundary conditions are imposed on the circular part of the boundary such that (12.1) remains the exact solution.

One can also alter the domain, and the severity of the singularity, using the parameter $NTRI$, where $1 \leq NTRI \leq 8$. If $NTRI = 8$ the entire circle is used as the domain; if $NTRI < 8$, only the first $NTRI$ triangles are used. Some examples are shown in figure 12.1

Boundary conditions are chosen on the outer boundary such that (12.1) remains the true solution, but now α varies as

$$\alpha = \frac{2}{NTRI}$$

For $1 \leq NTRI \leq 2$, the solution is smooth; for $3 \leq NTRI \leq 8$, the solution is singular, but the strength of the singularity varies with the number of triangles in the initial triangulation.

Since the exact solution is known, we can compute the exact error. For this test problem, the function QXY is defined to be the exact error for $ITYPE = 1$, the error in the gradient for $ITYPE = 2$ and $ITYPE = 3$, and the true solution for $ITYPE = 4$. It is thus possible to graph the true error and its gradient using $TRIPLT$.

12.2. Test Problem SQUARE

In this test problem, a complicated equation is solved on a simple domain. The domain is always the unit square; boundary conditions on each side of the square can be independently specified as Dirichlet or natural. The region is specified as the union of eight right triangles, so $TRIGEN$ should not be called to generate an initial mesh.

The coefficient functions are defined by

$$\begin{aligned}
a_1 &= A1X\,\frac{\partial u}{\partial x} + A1Y\,\frac{\partial u}{\partial y} + A1U\,u \\
a_2 &= A2X\,\frac{\partial u}{\partial x} + A2Y\,\frac{\partial u}{\partial y} + A2U\,u \\
f &= -BUX\,\frac{\partial u}{\partial x} - BUY\,\frac{\partial u}{\partial y} - CU0 - CU1\,u \\
&\quad -CU2\,u^2 - CU3\,u^3 - CEXP\,e^u \\
g_1 &= -EU0 \\
g_2 &= -DU0 - DU1\,u
\end{aligned}$$

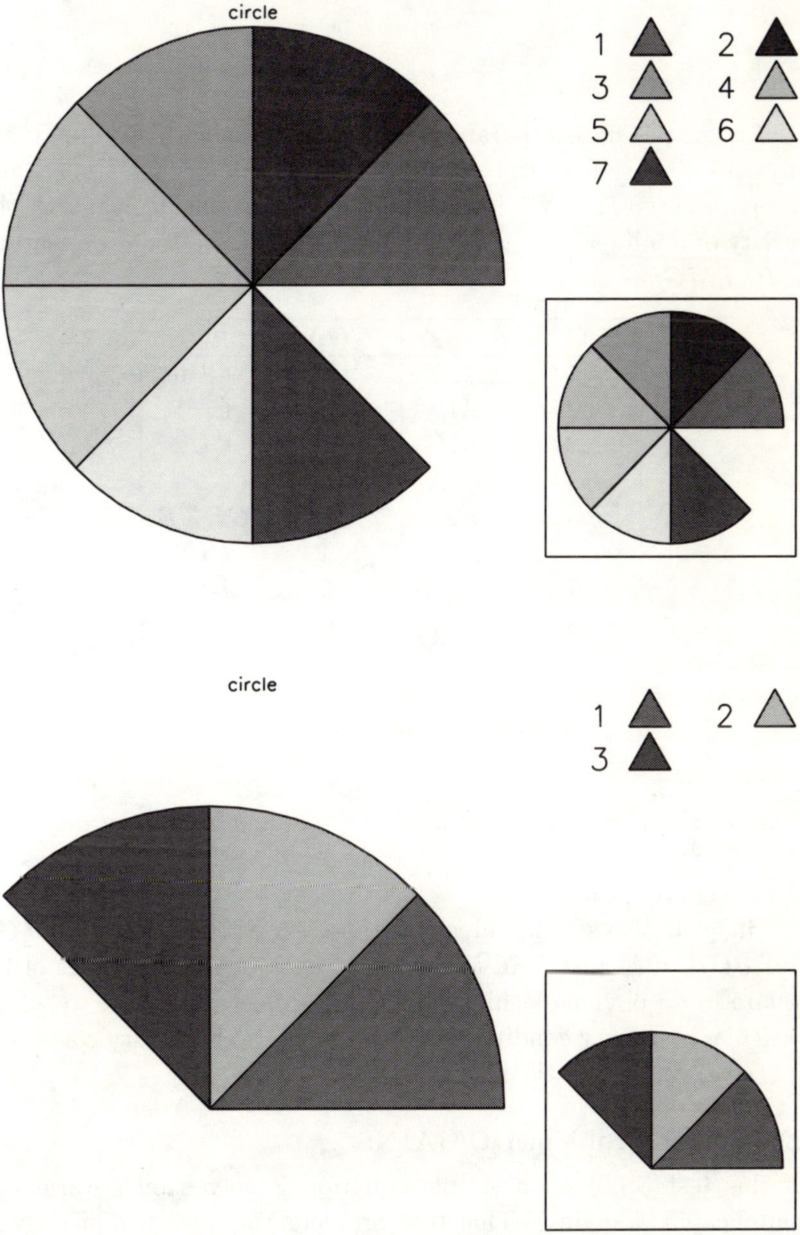

Fig. 12.1. *The domain for* $NTRI = 7$ *and* $NTRI = 3$.

and the functional ρ by

$$p_1 = u^2$$
$$p_2 = 0 \tag{12.2}$$

Any of the sixteen parameters can be set using *USRCMD*, and any can be used as the continuation parameter λ by specifying the parameter *ICONT* in *USRCMD* as in table 12.1. With this variety of nonlinearities, one can exercise most continuation features of *PLTMG*.

ICONT	λ	*ICONT*	λ
1	*A1X*	9	*CU0*
2	*A1Y*	10	*CU1*
3	*A1U*	11	*CU2*
4	*A2X*	12	*CU3*
5	*A2Y*	13	*CEXP*
6	*A2U*	14	*DU0*
7	*BUX*	15	*DU1*
8	*BUY*	16	*EU0*

Table 12.1.
Possible settings for ICONT.

If $ICONT = 0$, then none of the parameters is regarded as λ, and one should set $IPROB = 8$ to signify that the problem does not involve continuation.

One can also set the integer parameters *LEFT*, *RIGHT*, *TOP* and *BOTTOM* in *USRCMD*. These refer to the four sides of the square in an obvious fashion, and can be individually set to -1 for Dirichlet boundary conditions or $+1$ for natural boundary conditions for the given side of the square.

12.3. Test Problem DOMAINS

In this test problem, a simple equation is solved on a variety of complicated domains. This test problem was designed mainly to exercise *TRIGEN* and *SKELTN*.

The problem to be solved is the Poisson equation

$$-\Delta u = 1$$

with a combination of homogeneous Dirichlet and Neumann boundary conditions.

There is one parameter $DOMAIN$, satisfying $1 \leq DOMAIN \leq 6$, specifying the domain to be used. The parameter $DOMAIN$ is set in $USRCMD$. The various possibilities are shown in figures 12.2-12.4. All domains are defined by skeletons, so $TRIGEN$ must be called to generate an initial triangulation.

12.4. Test Problem NACA

Test problem $NACA$ solves the equation of potential flow in one of several domains. The equation is of the form

$$-\nabla \cdot \rho(\nabla u)\nabla u = 0$$

where

$$\rho(\nabla u) = (1 - u_x^2 - u_y^2)^{\frac{1}{\gamma-1}}$$

and $\gamma = 1.4$. The local mach number is computed in QXY and is given by

$$q = \sqrt{\frac{2c}{\gamma - 1}}$$

$$c = \frac{1}{1 - u_x^2 - u_y^2} - 1$$

There are three possible domains, chosen using the parameter $DOMAIN$ in $USRCMD$. The three domains are shown in figures 12.5-12.7. All regions are defined by skeletons, so $TRIGEN$ must be used to generate an initial mesh.

For the cases $DOMAIN = 1$ and $DOMAIN = 2$, the interior holes are shaped as Naca0012 airfoils. In these cases, Neumann boundary conditions are imposed everywhere. There are two parameters in $USRCMD$ that affect these problems. The parameter $MINF$, specifying the mach number at infinity M_∞, is used to set the boundary conditions on the outer boundary, and is also the continuation parameter λ for these problems. The parameter $ANGLE$ specifies the angle of attack (in degrees).

When the local mach number is less than one the flow is subsonic; $PLTMG$ will work well in regions when the flow is entirely subsonic. As the M_∞ is increased, the solution will begin to develop regions of supersonic flow near the airfoils; $PLTMG$ will continue to work

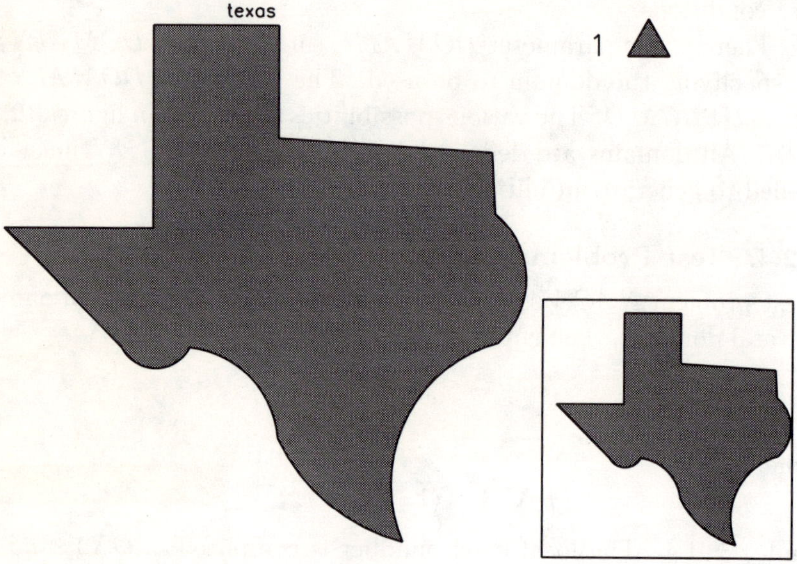

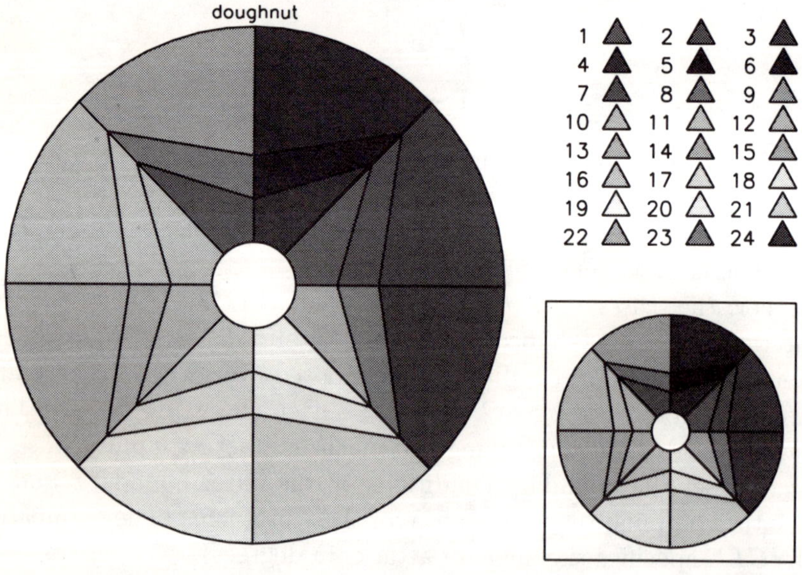

Fig. 12.2. *The domains for* $DOMAIN = 1$ *and* $DOMAIN = 2$.

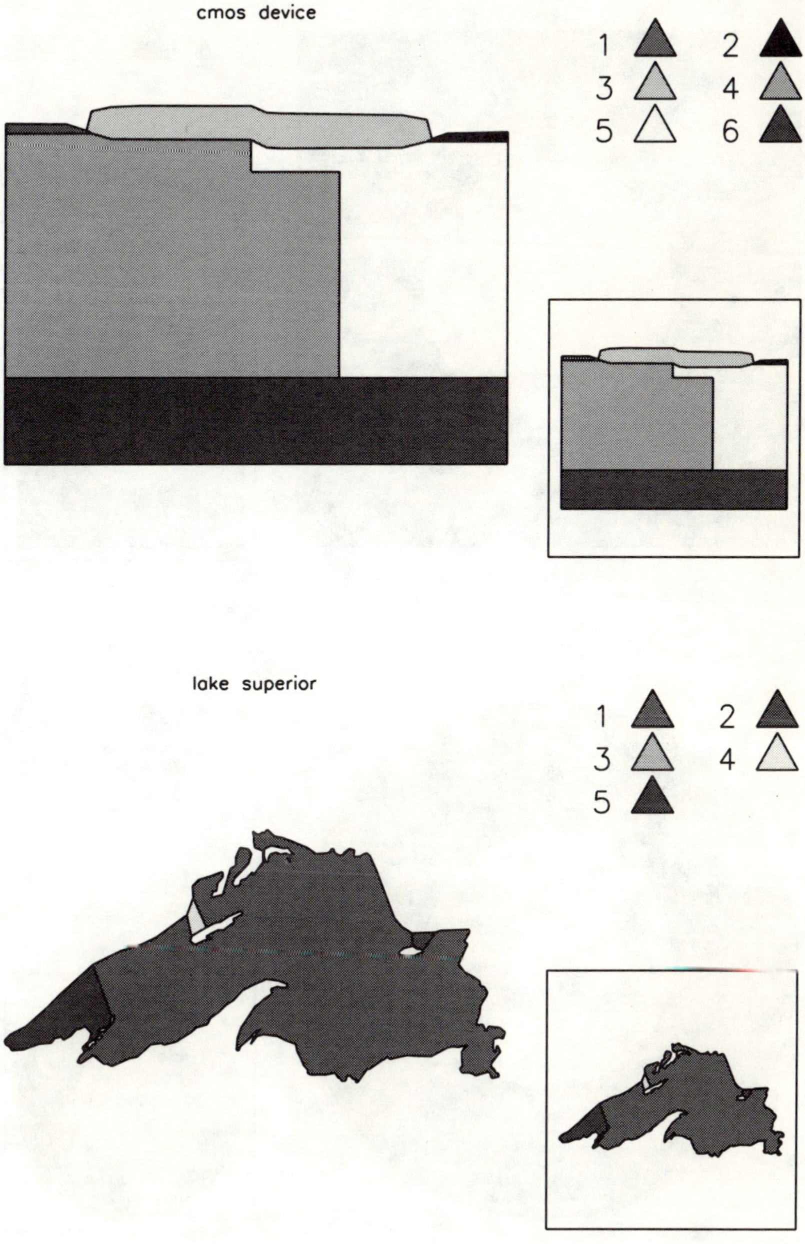

Fig. 12.3. *The domains for DOMAIN = 3 and DOMAIN = 4.*

Fig. 12.4. *The domains for* $DOMAIN = 5$ *and* $DOMAIN = 6$.

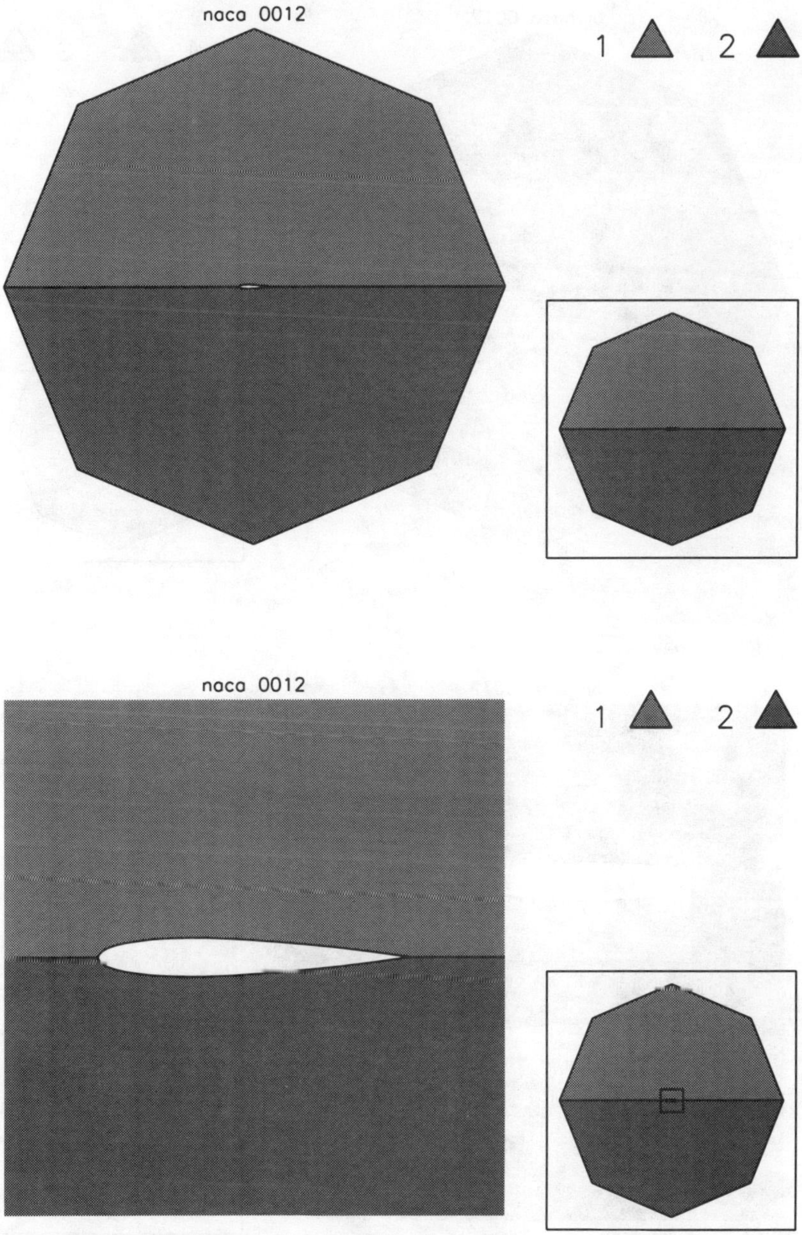

Fig. 12.5. *The domain for* $DOMAIN = 1$ *and a detail.*

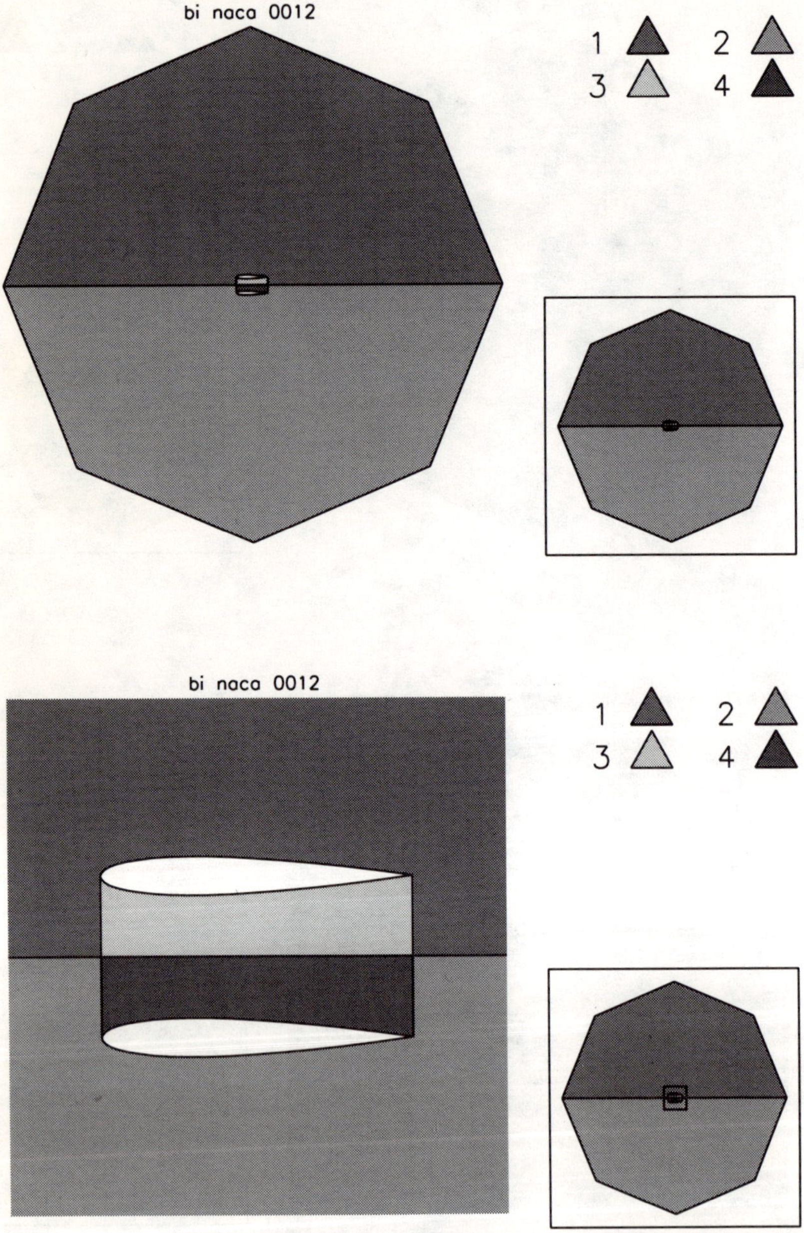

Fig. 12.6. *The domain for $DOMAIN = 2$ and a detail.*

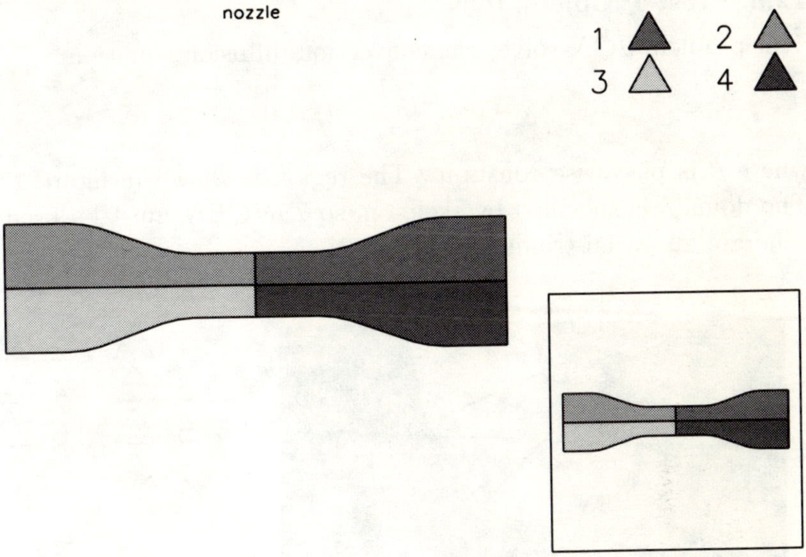

Fig. 12.7. *The domain for DOMAIN = 3.*

as these regions are forming, but eventually it will fail, as the underlying discretization used by *PLTMG* is not really appropriate for hyperbolic problems.

For the case *DOMAIN* = 3, the flow is through the nozzle shaped region. Homogeneous Dirichlet boundary conditions are imposed at the inlet (left side). Constant nonhomogeneous boundary conditions are imposed at the outlet (right side). This can be controlled by the user though the *USRCMD* parameter *RBC*, which is the continuation parameter λ for this problem. The right hand boundary condition is a constant multiple of *RBC*. The remaining boundary conditions are homogeneous Neumann. As with the other cases, when the flow is subsonic, *PLTMG* can handle the problem well. As *RBC* is increased, the solution will develop regions of supersonic flow. *PLTMG* will eventually fail when the supersonic region becomes sufficiently large. As with the other domains, this is due to an inappropriate discretization for hyperbolic problems.

12.5. Test Problem JCN

Test problem JCN solves the convection diffusion equation

$$-\nabla \cdot (\nabla u + \beta u) = 0$$

where β is piecewise constant. The region is shown in figure 12.8. The domain is specified by skeleton, so $TRIGEN$ must be used to generate an initial triangulation.

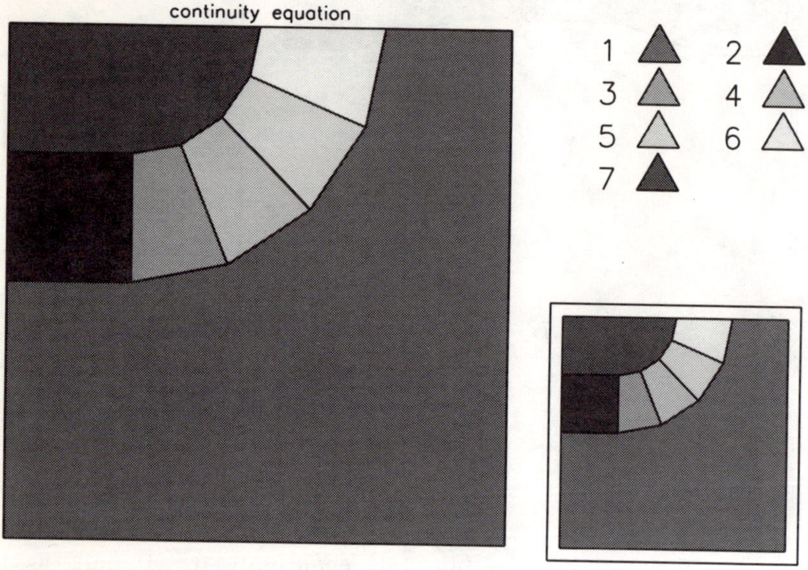

Fig. 12.8. *The domain for test problem JCN.*

This problem is an idealized model of the current continuity equation from the semiconductor device model that we have used to study the stability of discretizations used in device simulation. The problem has seven regions; $\beta = 0$ in regions one and seven. In the other five regions it has a magnitude of approximately 10^4, and is directed radially in each of the five subregions. The solution develops steep gradients at the junction between region seven and the five adjoining subregions.

Constant nonhomogeneous Dirichlet boundary conditions are specified along the bottom of the domain, and on the left hand portion of the top of the domain. Homogeneous Neumann boundary conditions are imposed elsewhere. The parameters TOP and

$BOTTOM$ in $USRCMD$ can be used to reset the Dirichlet boundary conditions on the top and bottom of the domain. The parameter DU can be used to adjust the size of β in regions 2-5; in particular, the magnitude of β in these five regions is proportional to DU.

Our original purpose in constructing this example was to test the sensitivity of various upwinding techniques [6] to poor element geometries. Since the goal of $TRIGEN$ is to produce elements with good geometries, the $USRCMD$ for this problem includes a procedure for systematically degrading the quality of the triangulation by introducing new elements with obtuse angles. This is done by dividing a triangle into three new triangles by connecting its baricenter to its vertices. Repeated application of this procedure can produce triangulations with interior angles arbitrarily close to π.

The $USRCMD$ parameters $OBTUSE$ and ISW are used to specify the type of mesh modifications. After $TRIGEN$ has been used to make a triangulation, one can modify the triangulation in $USRCMD$ by setting $OBTUSE = 1$. Then if $ISW = 0$, all elements in the current input triangulation are divided into three as described above. If $ISW = I$, with $1 \leq I \leq NT$, then only triangle I in the current mesh is trisected. If $ISW = -I$, with $1 \leq I \leq 7$, then all elements in region I are refined. An example is shown in figure 12.9.

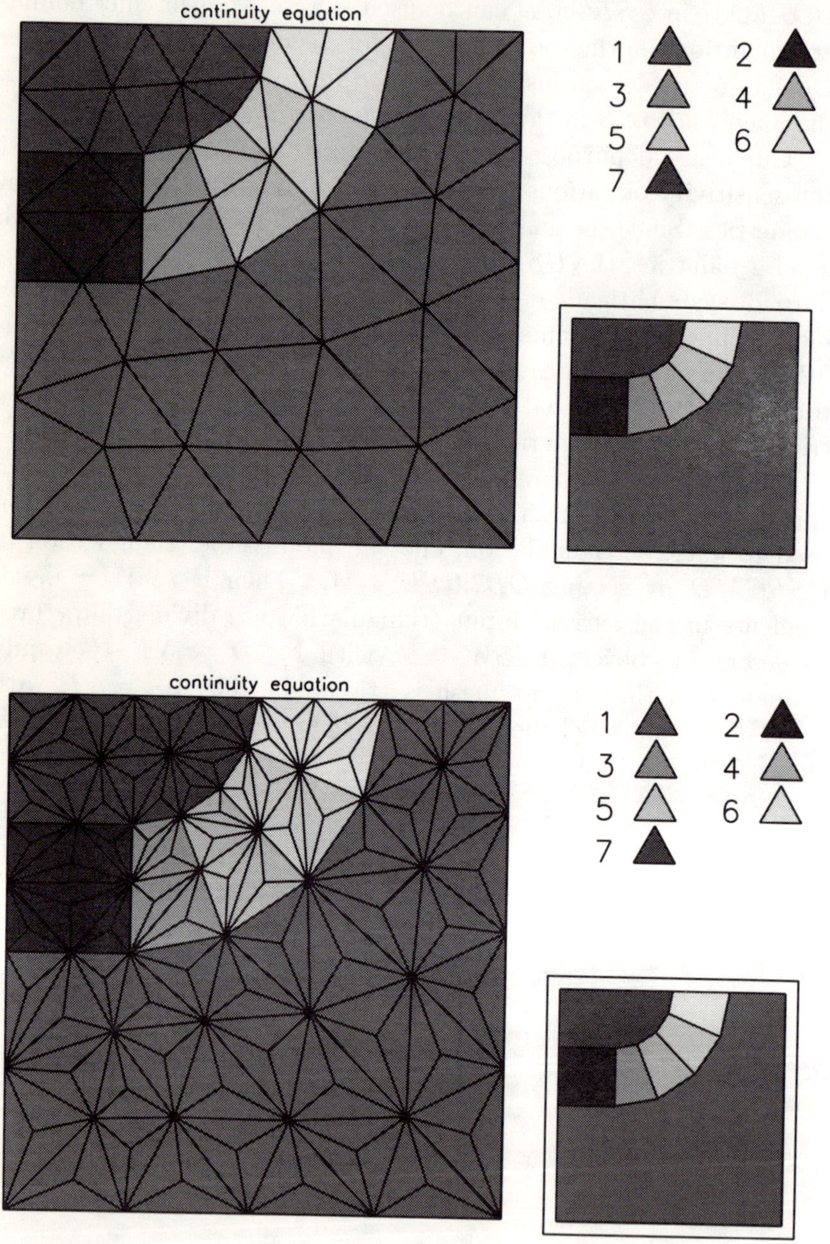

Fig. 12.9. *A triangulation produced by* $TRIGEN$ *(right), and the corresponding triangulation after a call to* $USRCMD$ *with* $OBTUSE = 1$ *and* $ISW = 0$.

References

[1] I. BABUŠKA, *Feedback, adaptivity, and a-posteriori estimates in finite elements: aims, theory, and experience*, in Accuracy Estimates and Adaptivity for Finite Elements, J. Wiley and Sons, New York, 1986.

[2] I. BABUŠKA AND A. K. NOOR, *Quality assessment and control of finite element solutions*, Tech. Report BN-1049, Institute for Physical Science and Technology, University of Maryland, 1986.

[3] I. BABUŠKA AND W. C. RHEINBOLDT, *Error estimates for adaptive finite element computations*, SIAM J. Numer. Anal., 15 (1978), pp. 736–754.

[4] R. E. BANK, *Analysis of a local a-posteriori error estimate for elliptic equations*, in Accuracy Estimates and Adaptivity for Finite Elements, J. Wiley and Sons, New York, 1986, pp. 119–128.

[5] ——, *A posteriori error estimates, adaptive local mesh refinement, and multigrid iteration*, in Multigrid Methods II: Proceedings, Cologne 1985 (Lecture Notes in Mathematics 1228), Springer-Verlag, Heidelberg, 1986, pp. 7–23.

[6] R. E. BANK, J. F. BÜRGLER, W. FICHTNER, AND R. K. SMITH, *Some upwinding techniques for finite element approximations of convection-diffusion equations*, Numer. Math.

[7] R. E. BANK AND T. F. CHAN, *PLTMGC: A multi grid continuation program for parameterized nonlinear elliptic systems*, SIAM J. Sci. Stat. Comput., 7 (1986), pp. 540–559.

[8] R. E. BANK AND T. F. DUPONT, *Analysis of a two level scheme for solving finite element equations*, Tech. Report CNA-159, Center for Numerical Analysis, University of Texas at Austin, 1980.

[9] R. E. BANK, T. F. DUPONT, AND H. YSERENTANT, *The hierarchical basis multigrid method*, Numer. Math., 52 (1988), pp. 427–458.

[10] R. E. BANK AND H. D. MITTELMANN, *Continuation and multigrid for nonlinear elliptic systems*, in Multigrid Methods II: Proceedings,

Cologne 1985 (Lecture Notes in Mathematics 1228), Springer-Verlag, Heidelberg, 1986, pp. 24–38.

[11] ——, *Stepsize selection in continuation procedures and damped Newton's method*, J. Comp. and Appl. Math., 26 (1989), pp. 67–78.

[12] R. E. Bank and D. J. Rose, *Global approximate Newton methods*, Numer. Math., 37 (1981), pp. 279–295.

[13] ——, *A multi-level Newton method for nonlinear finite element equations*, Math. Comp., 39 (1982), pp. 453–465.

[14] ——, *On the complexity of sparse Gaussian elimination via bordering*, SIAM J. Sci. Stat. Comput., 11 (1990), pp. 145–160.

[15] R. E. Bank, A. H. Sherman, and A. Weiser, *Refinement algorithms and data structures for regular local mesh refinement*, in Scientific Computing, Imacs/North Holland, 1983, pp. 3–17.

[16] R. E. Bank and R. K. Smith, *General sparse elimination requires no permanent integer storage*, SIAM J. Sci. Stat. Comput., 8 (1987), pp. 574–584.

[17] R. E. Bank and A. Weiser, *Some a-posteriori error estimators for elliptic partial differential equations*, Math. Comp., 44 (1985), pp. 283–301.

[18] R. E. Bank and H. Yserentant, *Some remarks on the hierarchical basis multigrid method*, in Domain Decomposition Methods, SIAM, Philadelphia, 1988, pp. 140–146.

[19] T. F. Chan, *Deflation techniques and block-elimination algorithms for solving bordered singular systems*, SIAM J. Sci. Stat. Comput., 5 (1984), pp. 121–134.

[20] J. J. Dongarra and E. Grosse, *Distribution of mathematical software via electronic mail*, Comm. ACM, 30 (1987), pp. 403–407.

[21] A. George and J. Liu, *Computer Solution of Large Sparse Positive Definite Systems*, Prentice-Hall, Englewood Cliffs, NJ, 1981.

[22] H. D. Mittelmann and H. Weber, *Multigrid solution of bifurcation problems*, SIAM J. Sci. Stat. Comp., 6 (1985), pp. 49–60.

[23] W. C. Rheinboldt, *Error estimates and adaptive techniques for nonlinear parameterized equations*, in Accuracy Estimates and Adaptivity for Finite Elements, J. Wiley and Sons, New York, 1986, pp. 163–180.

[24] ——, *Error questions in the computation of solution manifolds of parameterized equations*, Tech. Report ICMA-87-112, Institute for Computational Mathematics and Applications, University of Pittsburgh, 1987.

[25] D. J. Rose, *A graph theoretic study of the numeric solution of sparse positive definite systems*, in Graph Theory and Computing, Academic Press, New York, 1972.

[26] A. WEISER, *Local-Mesh, Local-Order Adaptive Finite Element Methods with A-Posteriori Error Estimators for Elliptic Partial Differential Equations*, PhD thesis, Yale University, 1981.

[27] H. YSERENTANT, *Hierarchical bases of finite element spaces in the discretization of nonsymmetric elliptic boundary value problems*, Computing, 35 (1985), pp. 39–49.

[28] ——, *On the multi-level splitting of finite element spaces for indefinite elliptic boundary value problems*, SIAM J. Numer. Anal., 23 (1986), pp. 581–595.

Index